An Introduction to Acceptance Sampling and SPC with R

An Introduction to Acceptance Sampling and SPC with R

John Lawson

CRC Press

Taylor & Francis Group

Boca Raton London New York

CRC Press is an imprint of the
Taylor & Francis Group, an **informa** business

First edition published [2021]
by CRC Press
6000 Broken Sound Parkway NW, Suite 300, Boca Raton, FL 33487-2742

and by CRC Press
2 Park Square, Milton Park, Abingdon, Oxon, OX14 4RN

ISBN: 9780367569952 (hbk)
ISBN: 9780367555764 (pbk)
ISBN: 9781003100270 (ebk)

Typeset in Computer Modern font
by KnowledgeWorks Global Ltd.

Access the Support Material: https://www.routledge.com/9780367555764

Contents

Preface

This book is an introduction to statistical methods used in monitoring, controlling and improving quality. Topics covered are: acceptance sampling; Shewhart control charts for Phase I studies; graphical and statistical tools for discovering and eliminating the cause of out-of-control-conditions; Cusum and EWMA control charts for Phase II process monitoring; and the design and analysis of experiments for process troubleshooting and discovering ways to improve process output. Origins of statistical quality control is presented in Chapter 1, and the technical topics presented in the remainder of the book are those recommended in the ANSI/ASQ/ISO guidelines and standards for industry. The final chapter combines everything together by discussing modern management philosophies that encourage the use of the technical methods presented earlier.

This would be a suitable book for an one-semester undergraduate course emphasizing statistical quality control for engineering majors (such as manufacturing engineering or industrial engineering), or a supplemental text for a graduate engineering course that included quality control topics.

A unique feature of this book is that it can be read free online at https://bookdown.org/home/authors/ by scrolling to the author's name (John Lawson). A physical copy of the book will also available from CRC Press.

Prerequisites for students would be an introductory course in basic statistics with a calculus prerequisite and some experience with personal computers and basic programming. For students wanting a review of basic statistics and probability, the book *Introduction to Probability and Statistics Using R* by Kerns[52] is available free online at (https://archive.org/details/IPSUR). A review of Chapters 3 (Data Description), 4 (Probability), 5 (Discrete Distributions), 6 (Continuous Distributions), 8 (Sampling Distributions), 9 (Estimation), 10 (Hypothesis Testing), and 11 (Simple Linear Regression) will provide adequate preparation for this book.

In this book, emphasis is placed on using computer software for calculations. However, unlike most other quality control textbooks that illustrate the use of hand calculations and commercial software, this book illustrates the use of the open source R software (like books by Cano et. al.[17] and Cano et. al.[16]). Kerns's book (Kerns 2011), mentioned above, also illustrates the use of R for probability and statistical calculations. Data sets from standard quality control textbooks are used as examples throughout this book so that the interested reader can compare the output from R to the hand calculations or output of commercial software shown in the other books.

As an open source high-level programming language with flexible graphical output options, R runs on Windows, Mac and Linux operating systems, and has add-on packages that equal or exceed the capabilities of commercial software for statistical methods used in quality control. It can be downloaded free from the Comprehensive R Archive Network (CRAN) https://cran.r-project.org/. The RStudio Integrated Development Environment (IDE) provides a command interface and GUI. A basic tutorial on RStudio is available at http://web.cs.ucla.edu/ gulzar/rstudio/basic-tutorial.html, also Chester Ismay's book *Getting used to R, RStudio, and R* provides a more details for new users of R and RStudio. It can also be read online free by scrolling down to the author's name (Chester Ismay) on https://bookdown.org/home/authors/, and a pdf version is available. RStudio can be downloaded from https://www.rstudio.com/products/rstudio/download/. Instructions for installing R and RStudio on Windows, Mac and Linux operating systems can be found at https://socserv.mcmaster.ca/jfox/Courses/ R/ICPSR/R-install-instructions.html.

The R packages illustrated in this book are AcceptanceSampling[54], AQLSchemes[59], daewr[58], DoE.Base[33], FrF2[34], IAcsSPCR[60], qcc[85], qualityTools[81], spc[55], spcadjust[29], SixSigma[15], and qicharts[3]. At the time of this writing, all the R packages illustrated in this book are available on the Comprehensive R Archive Network except IAcsSPCR. It can be installed from Rforge with the command install.packages("IAcsSPCR", repos="http://R-Forge.R-project.org"). The latest version (3.0) of the qcc package is illustrated in this book. The input and output of qcc version 2.7 (that is on CRAN) is slightly different. The latest version (3.0) of qcc can be installed from GitHub at https://luca-scr.github.io/qcc/https://luca-scr.github.io/qcc/.

For students with no experience with R, an article giving a basic introduction to R can be found at https://www.red-gate.com/simple-talk/dotnet/software-tools/r-basics/. Chapter 2 of *Introduction to Probability and Statistics using R* by Kerns[52] is also an introduction to R, as well as the pdf book *R for Beginners* by Emmanuel Paradis that can be downloaded from https://cran.r-project.org/doc/contrib/Paradis-rdebuts_en.pdf.

All the R code illustrated in the book plus additional resources can be downloaded from https://lawsonjsl7.netlify.app/webbook/ by clicking on the cover of this book, and then on the link for code in a particular chapter.

Acknowledgements Many thanks to suggestions for improvements given by authors of R packages illustrated in this book, namely Andreas Kiemeier, author the Acceptance Sampling package, Luca Scrucca, author of the qcc package, and Ulrike Groemping, author of the DoE.Base and FrF2 packages. Also thanks to suggestions from students in my class and editing help from my wife Dr. Francesca Lawson.

About the author John Lawson is a Professor Emeritus from the Statistics Department at Brigham Young University where he taught from 1986 to 2019. He is an ASQ-CQE and he has a Masters Degree in Statistics from Rutgers University and a PhD in Applied Statistics from the Polytechnic Institute of N.Y. He worked as a statistician for Johnson & Johnson Corporation from 1971 to 1976, and he worked at FMC Corporation Chemical Division from 1976 to 1986 where he was the Manager of Statistical Services. He is the author of *Design and Analysis of Experiments with R*, CRC.Press, and the co-author (with John Erjavec) of *Basic Experimental Strategies and Data Analysis for Science and Engineering*, CRC Press.

List of Figures

List of Tables

1

Introduction and Historical Background

Statistical Quality Control includes both (1) the application of statistical sampling theory that deals with quality assurance and (2) the use of statistical techniques to monitor and control a process. The former includes acceptance-sampling procedures for inspecting incoming parts or raw materials, and the latter (often referred to as statistical process control or SPC) employs the use of control charts, continuous improvement tools, and the design of experiments for early detection and prevention of problems, rather than correction of problems that have already occurred.

1.1 Origins of Statistical Quality Control

Quality control is as old as industry itself, but the application of statistical theory to quality control is relatively recent. When AT&T was developing a nationwide telephone system at the beginning of the 20th century, sampling inspection was used in some form at the Western Electric Company (the AT&T equipment manufacturing arm). Also, according to an article in the *General Electric Review* in 1922, some formal attempts at scientific acceptance sampling techniques were being made at the G.E. Lamp Works.

At Western Electric, an Inspection Engineering Department was formed, which later became the Quality Assurance Department of the Bell Telephone Laboratories. In 1924, Walter Shewhart, a physicist and self-made statistician, was assigned to examine and interpret inspection data from the Western Electric Company Hawthorn Works. It was apparent to him that little useful inference to the future could be drawn from the records of past inspection data, but he realized that something serious should be done and he conceived the idea of statistical control. It was based on the premise that no action can be repeated exactly. Therefore, all manufactured product is subject to a certain amount of variation that can be attributed to a system of chance causes. Stable variation within this system is inevitable. However, the reasons for special cause variation outside this stable pattern can (and should) be recognized and eliminated.

The control chart he perceived was founded on sampling during production rather than waiting until the end of the production run. Action limits were calculated from the chance cause variation in the sample data, and the process could be immediately stopped and adjusted when additional samples fell outside the action limits. In that way, production output could be expected to stay within defined limits.

The early dissemination of these ideas was limited to the circulation of memos within the Bell Telephone System. However, the soundness of the proposed methods was thoroughly validated by staff at Western Electric and the Bell Telephone Laboratories. The methods worked effectively and were soon made part of the regular procedures of the production divisions. Shewhart's ideas were eventually published in his 1931 book *The Economic Control of Quality of Manufactured Product* [86].

W. Edwards Deming from the U.S. Department of Agriculture and the Census Bureau, who developed the sampling techniques that were first used in the 1940 U.S. Census, was introduced to Shewhart in 1927. He found great inspiration in Shewhart's theory of chance (that he renamed common) and special causes of variation. He realized these ideas could be applied not only to manufacturing processes but also to administrative processes by which enterprises are led and managed. However, many years later in a videotaped lecture Deming said that while Shewhart was brilliant, he made things appear much more difficult than necessary. He therefore spent a great deal of time copying Shewhart's ideas and devising simpler and more easily understood ways of presenting them.

Although Shewhart's control charts were effective in helping organizations control the quality of their own manufacturing processes, they were still dependent on the quality of raw materials, purchased parts and the prevailing quality control practices of their suppliers. For these reasons, sampling inspection of incoming parts remained an important part of statistical quality control.

Harold F. Dodge joined Western Electric Corporation shortly after Shewhart did. He wondered, "how many samples were necessary when inspecting a lot of materials", and began developing sampling inspection tables. When he was joined by Harry G. Romig, together they developed double sampling plans to reduce the average sample size required, and by 1927 they had developed tables for rectification inspection indexed by the lot tolerance and AOQL (average outgoing quality level). Rectification sampling required removal of defective items through 100% inspection of lots in which the number defective in the sample was too high. Dodge and Romig's sampling tables were published in the *Bell System Technical Journal* in 1941 [22].

The work of Shewhart, Dodge and Romig at Bell Telephone constituted much of the statistical theory of quality control at that time. In the 1930s, the Bell System engineers who developed these methods sought to popularize them in cooperation with the American Society for Testing and Materials, the American Standards Association, and the American Society of Mechanical Engineers. Shewhart also traveled to London where he met with prominent British statisticians and engineers.

Despite attempts to publicize them, adoption of statistical quality control in the United States was slow. Most engineers felt their particular situation was different and there were few industrial statisticians who were adequately trained in the new methods. By 1937, only a dozen or more mass production industries had implemented the methods in normal operation. There was much more rapid progress in Britain, however. There, statistical quality control was being applied to products such as coal, coke, textiles, spectacle glass, lamps, building materials, and manufactured chemicals (see Freeman[28]).

1.2   Expansion and Development of Statistical Quality Control during WW II

The initial reluctance to adopt statistical quality control in the United States was quickly overcome at the beginning of World War II. Manufacturing firms switched from the production of consumer goods to defense equipment. With the buildup of military personnel and material, the armed services became large consumers of American goods, and they had a large influence on quality standards.

The military had impact on the adoption of statistical quality control methods by industry in two different ways. The first was the fact that the armed services themselves adopted statistically derived sampling and inspection methods in their procurement efforts. The second was the establishment of a widespread educational program for industrial personnel at the request of the War Department.

Sampling techniques were used at the Picatinney Arsenal as early as 1934 under direction of L. E. Simon. In 1936, the Bell Telephone Laboratories were invited to cooperate with the Army Ordnance Department and the American Standards Association Technical Committee in developing war standards for quality control. In 1942, Dodge and Romig completed the Army Ordnance Standard Sampling Inspection Tables, and the use of these tables was introduced to the armed services through a number of intensive training courses.

The Ordnance Sampling Inspection Tables employed a sampling scheme based on an acceptable quality level (AQL). The scheme assumed that there would be a continuing stream of lots submitted by a supplier. If the supplier's quality level was worse than the AQL, the scheme would automatically switch to tightened inspection and the supplier would be forced to bear the cost of a high proportion of lots rejected and returned. This scheme encouraged suppliers to improve quality.

In 1940, the military established a widespread training program for industrial personnel, most notably suppliers of military equipment. At the request of the War Department, the American Standards Association developed American War Standards Z1.1-1941 and Guide for Quality Control Z.1-2-1941, Control Chart Method of Analyzing Data–1941, and the Control Chart

Method of Controlling Quality during Production Z1.3-1942. These defined the American control chart practice and were used as the text material for subsequent training courses that were developed at Stanford by Holbrook Working, E. L. Grant, and W. Edwards Deming. In 1942 this intensive course on statistical quality control was given at Stanford University to representatives of the war industries and procurement agencies of the armed services.

The early educational program was a success. That success, along with the suggestion from Dr. Walter Shewhart that Federal assistance should be given to American war industries in developing applications of statistical quality control, led the Office of Production and Research and Development (OPRD) of the War Production Board to establish a nationwide program. The program combined assistance in developing intensive courses for high ranking executives from war industry suppliers and direct assistance to establishments on specific quality control problems. The specific needs to be addressed by this program for the war-time development of statistical quality control were:

1. Education of industrial executives regarding the basic concepts and benefits of statistical quality control.

2. Training of key quality control personnel in industry.

3. Advisory assistance on specific quality control problems.

4. Training of subordinate quality control personnel.

5. Training of instructors.

6. Publication of literature.

The training of instructors was regarded as an essential responsibility of the OPRD program. The instructors used were competent and experienced university teachers of statistics who only needed to (1) extend their knowledge in in the specific techniques and theory most relevant to statistical quality control, (2) become familiar with practical applications, and (3) learn the instructional techniques that had been found to be most useful.

· The plan was to have courses for key quality control personnel from industry given at local educational institutions, which would provide an instructor from their own staff. This plan was implemented with administrative assistance and grants from the Engineering, Science and Management War Training Program (ESMWT) funded by the U.S. Office of Education.

Much of the training of subordinate quality control personnel was given in their own plants by those previously trained. To stimulate people to actively advance their own education, the OPRD encouraged local groups to form. That way neighboring establishments could exchange information and experiences. These local groups resulted in the establishment of many regional quality control societies. The need for literature on statistical quality control was satisfied by publications of the American Standards Association and articles in engineering and technical journals.

As a result of all the training and literature, statistical quality control techniques were widely used during the war years. They were instrumental in ensuring the quality and cost effectiveness of manufactured goods as the nations factories made the great turnaround from civilian to military production. For example, military aircraft production totaled 6000 in 1940 and jumped to 85,000 in 1943. Joseph Stalin stated that without the American production, the Allies could never have won the war.

At the conclusion of the War in 1946, seventeen of the local quality control societies formed during the war organized themselves into the American Society for Quality Control (ASQC). This society has recently been renamed the American Society for Quality (ASQ) to reflect the fact that Quality is essential to much more than manufacturing firms. It is interesting to note that outside the board room of ASQ in Milwaukee Wisconsin stands an exhibit memorializing W.E. Deming's famous Red Bead Experiment[1] (a teaching tool) that was used during the war effort to show managers the futility of the standard reaction to common causes of variation.

The development and use of sampling tables and sampling schemes for military procurement continued after the war, resulting in the MIL-STD 105A attributes sampling scheme, which was later revised as 105B, 105C, 105D, and 105E. In addition, variables sampling schemes were developed that eventually resulted in MIL-STD 414.

1.3 Use and Further Development of Statistical Quality Control in Post-War Japan

After the war, companies that had been producing defense equipment resumed producing goods for public consumption. Unfortunately, the widespread use of statistical quality control methods that had been used so effectively in producing defense equipment did not carry over into the manufacture of civilian goods. Women who filled many positions in inspection and quality improvement departments during the war left the workforce and were replaced by military veterans who were not trained in the vision and technical use of SPC. Industry in Europe lay in the ruins of war, and the overseas and domestic demand for American manufactured goods exceeded the supply. Because they could sell everything they produced, company top managers failed to see the benefits of the extra effort required to improve quality. As the U. S. economy grew in the 1950s there were speculations of a coming recession, but it never came. The demand for products continued to increase leaving managers to believe they were doing everything right.

---

[1]This was used to teach the futility of reacting to common cause variation by making changes to the process, and the need for managers to understand the system before placing unreasonable piecework standards upon workers—see https://www.youtube.com/watch?v=ckBfbvOXDvU

At the same time, U.S. occupation forces were in Japan trying to help rebuild their shattered economy. At the request of General Douglas McArthur, W. E. Deming was summoned to help in planning the 1951 Japan Census. Deming's expertise in quality control techniques and his compassion for the plight of the Japanese citizens brought him an invitation by the Japanese Society of Scientists and Engineers (JUSE) to speak with them about SPC. At that time, what was left of Japanese manufacturing was almost worse than nothing at all. The label Made in Japan was synonymous with cheap junk in worldwide markets.

Members of JUSE had been interested in Shewhart's ideas and they sought an expert to help them understand how they could apply them in the reconstruction of their industries. At their request, Deming, a master teacher, trained hundreds of Japanese academics, engineers, and managers in statistical quality control techniques. However, he was troubled by his experience in the United States where these techniques were only widely used for a short time during the war years. After much speculation, Deming had come to the conclusion that in order for the use of these statistical techniques to endure, a reputable and viable management philosophy was needed that was consistent with statistical methods. As a result Deming developed a philosophy that he called the 14 Points for Management. There were originally less than 14 points, but they evolved into 14 with later experience.

Therefore, when Deming was invited by JUSE to speak with them about SPC, he agreed to do so only if he could first talk directly to top management of companies. In meetings with executives in 1950, his main message was his Points for Management[2] and the following simple principle: (1) Improve

---

[2]https://asq.org/quality-resources/total-quality-management/deming-points

1. Create constancy of purpose for improving products and services.
2. Adopt the new philosophy.
3. Cease dependence on inspection to achieve quality.
4. End the practice of awarding business on price alone; instead, minimize total cost by working with a single supplier.
5. Improve constantly and forever every process for planning, production and service.
6. Institute training on the job.
7. Adopt and institute leadership.
8. Drive out fear.
9. Break down barriers between staff areas.
10. Eliminate slogans, exhortations and targets for the workforce.
11. Eliminate numerical quotas for the workforce and numerical goals for management.
12. Remove barriers that rob people of pride of workmanship, and eliminate the annual rating or merit system.
13. Institute a vigorous program of education and self-improvement for everyone.
14. Put everybody in the company to work accomplishing the transformation.

Quality $\Rightarrow$ (2) Less Rework and Waste $\Rightarrow$ (3) Productivity Improves $\Rightarrow$ (4) Capture the Market with Lower Price and Better Quality $\Rightarrow$ (5) Stay in Business $\Rightarrow$ (6) Provide Jobs.

With nothing to lose, the Japanese manufactures applied the philosophy and techniques espoused by Deming and other American experts in Quality. The improved quality, along with lower cost of the goods, enabled the Japanese to create new international markets for Japanese products, especially automobiles and consumer electronics. Japan rose from the ashes of war to becoming one of the largest economies in the world. When Deming declined to accept royalties on the published transcripts of his 1950 lectures, the JUSE board of directors used the proceeds to establish the Deming Prize, a silver medal engraved with Demings profile. It is given annually in a nationally televised ceremony to an individual for contributions in statistical theory and to a company for accomplishments in statistical application.

As Deming predicted in 1950, Japanese products gained respect in worldwide markets. In addition, Japanese began to contribute new insights to the body of knowledge regarding SQC. Kaoru Ishikawa, a Deming Prize recipient, developed the idea of quality circles where foreman and workers met together to learn problem solving tools and apply them to their own process. This was the beginning of participative management. Ishikawa also wrote books on quality control including his *Guide to Quality Control* that was translated to English and defined the 7 basic quality tools to be discussed later in Chapter 4. Genichi Taguchi developed the philosophy of off-line quality control where products and processes are designed to be insensitive to common sources of variation that are outside the design engineers control. An example of this will be shown in Chapter 5.

When the Arab Oil Embargo caused the price of oil to increase from $3 to $12 per barrel in 1973, it created even more demand for small fuel-efficient Japanese cars. In the U.S, when drivers began to switch to the smaller cars, they noticed that in addition to being more fuel efficient, they were more reliable and less problematic. By 1979, U.S. auto manufactures had lost a major share of their market, and many factories were closed and workers laid off. This was a painful time, and when the NBC Documentary "If Japan Can, Why Can't We" aired in 1979, it was instrumental in motivating industry leaders to start re-learning the quality technologies that that had helped Japan's manufacturing, but were in disuse in the U.S.

1.4   Re-emergence of Statistical Quality Control in U.S. and the World

Starting about 1980, top management of large U.S. Companies began to accept quality goals as one of the strategic parameters in business planning along with the traditional marketing and financial goals. For example, Ford Motor Company adopted the slogan "Quality is Job 1", and they followed the plan of the Defense department in WW II by setting up training programs for their own personnel and for their suppliers. Other companies followed suit, and the quality revolution began in the U.S.

Total Quality Management or TQM was adopted by many U.S companies. This management system can be described as customer focused and involves all employees in continual improvement efforts. As described on the American Society for Quality (ASQ) website[3] "It uses strategy, data, and effective communications to integrate the quality discipline into the culture and activities of the organization".

Based on efforts like this, market shares of U.S. companies rebounded in manufactured goods such as automobiles, electronics, and steel. In addition, the definition of Quality expanded from just meeting manufacturing specifications to pleasing the customer. The means of providing quality through the expanded definition of SQC as espoused by Deming and others was adopted in diverse areas such as utility companies, health care institutions, banking and other customer service organizations.

A worldwide movement began using the same philosophy. In 1987, the International Standardization Organization created the Standards for Quality Assurance Systems (ISO 9000). It was the acknowledgement of the worldwide acceptance of the systems approach to producing Quality. ISO 9001 deals with the requirements that organizations wishing to meet the standard must fulfil. Certification of compliance to these standards are required for companies to participate in the European Free Trade Association (EFTA). Table 1.1 shows the number of companies ISO 9001 registered by country.

TABLE 1.1

ISO 9001 Registrations by Country 2014

Rank	Country	ISO 9001 Registrations
1	China	342,180
2	Italy	168,960
3	Germany	55,563
4	Japan	45,785
5	India	41,016
6	United Kingdom	40,200
7	Spain	36,005
8	United States	33,008
9	France	29,122
10	Austria	19,731

In 1988, the U.S. Congress established the Malcolm Baldrige National Quality Award named after the late secretary of commerce. This is similar to the Deming award in Japan, and it was a recognition by the U.S. government

---

[3]https://asq.org/quality-resources/total-quality-management

of the need to focus on the quality of products and services in order to keep the U.S. economy competitive.

Other changes to statistical quality control practices have also taken place. The U. S. Department of Defense discontinued support of their Military Standards for Sampling Inspection in order to utilize civilian standards as a cost savings. The ANSI/ASQ Z1.4 is the civilian standard in the U.S. that replaces the MIL-STD 105E Attribute Sampling Inspection Tables. It is best used for domestic transactions or in-house use. The ISO 2851-1 is the international standard. It reflects the current state of the art and is recommended for international trade, where it is often required. The U.S. civilian standard to replace MIL-STD 414 Variables Sampling Plans is ANSI/ASQ Z1.9. It was designed to make the inspection levels coincide with the Z1.4 plans for attributes and adopts common switching rules. ISO 3951-1 is the international version with plans closely matched to the ISO 2851-attribute plans. It is also used in international trade.

Another change in the application of technical methodologies for quality control and quality improvement is the use of the computer. Prior to 1963, the only tools available to engineers and statisticians for calculations were slide rules or mechanical or electro-mechanical calculators. Sampling inspection tables and Shewhart's control charts were developed with this fact in mind.

After availability of computers, software began to be developed for statistical calculations and SQC applications. However, much of the training materials and textbooks that have been developed since the 1980 comeback of SQC in U.S. industry still illustrates the techniques that can easily be implemented with hand calculations. However, this book will emphasize the use of the computer for SQC calculations.

Popular commercial software used in industry includes programs such as SynergySPC and SQCpack that can share data and reports between different computers. Others like SAS, Minitab 18, or StatGraphics Centurion combine SQC calculations with data manipulation features and a full suite of statistical analysis tools.

This book will illustrate the use of R, since it is a free programming language and environment for statistical computing. R was developed by Ross Ihaka and Robert Gentleman at the University of Auckland, New Zealand. It implements the S programming language that was developed at Bell Labs by John Chambers in 1976. R is highly extendible through functions and extensions. There are many user written packages for statistical quality control functions that are available on the Comprehensive Archive Network (CRAN).

2

Attribute Sampling Plans

2.1 Introduction

The quality and reliability of manufactured goods are highly dependent on the quality of component parts. If the quality of component parts is low, the quality and/or reliability of the end assembly will also be low. While some component parts are produced in house, many are procured from outside suppliers; the final quality is, therefore, highly dependent on suppliers.

In response to stiff competition, Ford Motor Company adopted procedural requirements for their suppliers in the early 1980s to insure the quality of incoming component parts. They demanded that all their suppliers show that their production processes were in a state of statistical control with a capability index greater than 1.5. Because Ford Motor Company bought such a large quantity of component parts from their suppliers, they were able to make this demand.

Smaller manufacturing companies may not have enough influence to make similar demands to their suppliers, and their component parts may come from several different suppliers and sub-contractors scattered across different countries and continents. However, by internal use of acceptance sampling procedures, they can be sure that the quality level of their incoming parts will be close to an agreed upon level. This chapter will illustrate how the `AcceptanceSampling` and `AQLSchemes` packages in R can be used both to create attribute sampling plans and sampling schemes and evaluate them.

2.2 Attribute Data

When numerical measurements are made on the features of component parts received from a supplier, quantitative data results. On the other hand, when only qualitative characteristics can be observed, attribute data results. If only attribute data is available, incoming parts can be classified as either conforming/nonconforming, non-defective/defective, pass/fail, or present/absent etc. Attribute data also results from inspection data (such as inspection of billing records), or the evaluation of the results of maintenance operations or

11

administrative procedures. Non conformance in these areas is also costly. Errors in billing records result in delayed payments and extra work to correct and resend the invoices. Non conformance in maintenance or administrative procedures, result in rework and less efficient operations.

2.3   Attribute Sampling Plans

A lot or Batch is defined as "a definite quantity of a product or material accumulated under conditions that are considered uniform for sampling purposes" [4]. The only way that a company can be sure that every item in an incoming lot of components from a supplier, or every one of their own records or results of administrative work completed, meets the accepted standard is through 100% inspection of every item in the lot. However, this may require more effort than necessary, and if the inspection is destructive or damaging, this approach cannot be used.

Alternatively, a sampling plan can be used. When using a sampling plan, only a random sample of the items in a lot is inspected. When the number of nonconforming items discovered in the sample of inspected components is too high, the lot is returned to the supplier (just as a customer would return a defective product to the store). When inspecting the records of administrative work completed, and the number of nonconforming records or nonconforming operations are too high in a lot or period of time, every item in that period may be inspected and the work redone if nonconforming. On the other hand, if the number of nonconforming items discovered in the sample is low, the lot is accepted without further concern.

Although a small manufacturing company may not be able to enforce procedural requirements upon their suppliers, the use of an acceptance sampling plan will motivate suppliers to meet the agreed upon acceptance quality level or improve their processes so that it can be met. Otherwise, they will have to accept returned lots which will be costly.

When inspecting only a random sample from a lot, there is always a nonzero probability that there are nonconforming items in the lot even when there are no nonconforming items discovered in the sample. However, if the customer and supplier can agree on the maximum proportion nonconforming items that may be allowed in a lot, then an attribute acceptance sampling plan can be used successfully to reject lots with proportion nonconforming above this level. The sampling plan can both maximize the probability of rejecting lots with a higher proportion nonconforming than the agreed upon level (benefit to the customer), and it can maximize the probability of accepting lots that have the proportion nonconforming at or below the agreed upon level (benefit to the supplier).

For a lot of N components, an attribute sampling plan consists of the number of items to be sampled, n, and the acceptance number or maximum number of nonconforming items, c, that can be discovered in the sample and still allow the lot to be accepted. The probability of accepting lots with varying proportions of nonconforming or defective items using an attribute acceptance sampling plan can be represented graphically by the OC (or Operating Characteristic) curve shown in Figure 2.1.

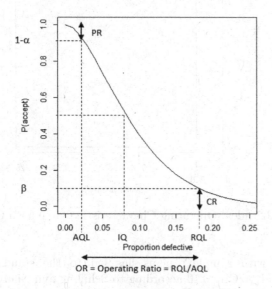

FIGURE 2.1: Operating Characteristic Curve

In this figure the AQL (Acceptance Quality Level) represents the agreed upon maximum proportion of nonconforming components in a lot. $1-\alpha$ represents the probability of accepting a lot with the AQL proportion nonconforming, and the PR=α is the producer's risk or probability that a lot with AQL proportion nonconforming is rejected. The IQ is the indifference quality level where 50% of the lots are rejected, and RQL is the rejectable quality level where there is only a small probability, β, of being accepted. The customer should decide on the RQL. CR=β is the customer's risk.

From the customer's point of view, a steeper OC curve with a smaller operating ratio (or ratio of the RQL to the AQL) is preferable. In this case, the probability of accepting any lot with greater than the AQL proportion nonconforming is reduced. To prevent rejected lots, the supplier will be motivated to send lots with the proportion nonconforming less than the AQL. The ideal OC curve is shown in Figure 2.2. It would result when 100% inspection is used or $n = N$ and c=AQL$\times N$. In this case, all lots with a proportion nonconforming (or defective) less than the AQL are accepted, and all lots with the proportion nonconforming greater than the AQL are rejected. As the fraction items sampled (n/N) in a sampling plan increases, the OC curve for that plan will approach the curve for the ideal case.

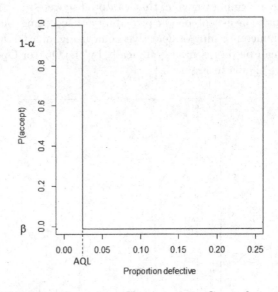

FIGURE 2.2: Ideal Operating Characteristic Curve for a Customer

Experience with sampling plans has led to the standard values of $\alpha=$PR$=0.05$ and $\beta=$CR$=0.10$ according to Schilling and Neubauer[84]. This will result in one lot in 20 rejected when proportion nonconforming is at the AQL, and only one lot in 10 accepted when the proportion nonconforming is at the RQL. When $\beta=0.10$, the RQL is usually referred to as the LTPD or Lot Tolerance Percent Defective. While $\alpha = .05$ and $\beta = 0.10$ are common, other values can be specified. For example, if nonconforming items are costly then it may be desirable to use a β less than 0.10, and if rejecting lots that have the AQL proportion nonconforming is costly, it may be desirable to use α less than 0.05.

2.4   Single Sample Plans

In a single sampling plan, as described in the last section, all n sample units are collected before inspection or testing starts. Single sampling plans specify the number of items to be sampled, n, and the acceptance number, c. Single sampling plans can be obtained from published tables such as MIL-STD-105E, ANSI/ASQ Standard Z1.4, ASTM International Standard E2234, and ISO Standard 2859. Plans in these published tables are indexed by the lot size and AQL. The tables are most useful for the case when a purchaser buys a continuing stream of lots or batches of components, and the purchaser and seller agree to use the tables. More about these published sampling plans will be discussed in Section 2.6.

Custom derived sampling plans can be constructed for inspecting isolated lots or batches. Analytic procedures have been developed for determining the sample size, n, and the acceptance number, c, such that the probability of accepting a lot with the AQL proportion nonconforming will be as close as possible to $1 - \alpha$, and the probability of accepting a lot with RQL proportion nonconforming will be as close as possible to β. These analytic procedures are available in the `find.plan()` function in the R package `AcceptanceSampling` that can be used for finding single sampling plans.

As an example of this function, consider finding a sampling plan where the AQL=0.05, α=.05, RQL $= 0.15$, and β=0.20 for a lot of 500 items. For a plan where the sample size is n and the acceptance number is c, the probability of accepting a lot of N=500 with D nonconforming or defective items is given by the cumulative Hypergeometric distribution, i.e.,

$$Pr(accept) = \sum_{i=0}^{c} \frac{\binom{D}{i}\binom{N-D}{n-i}}{\binom{N}{n}} \tag{2.1}$$

The `find.plan()` function attempts to find n and c so that the probability of accepting when $D = 0.05 \times N$ is as close to $1 - 0.05 = .95$ as possible, and the probability of accepting when $D = 0.15 \times N$ is as close to 0.20 as possible. The first statement in the R code below loads the `AcceptanceSampling` package. The next statement is the call to `find.plan()`. The first argument in the call, `PRP=c(0.05,0.95)`, specifies the producer risk point (AQL, $1-\alpha$); the second argument specifies the consumer risk point (RQL, β); the next argument, `type="hypergeom"`, specifies that the probability distribution is the hypergeometric; and the last argument, `N=500`, specifies the lot size.

```
R>library(AcceptanceSampling)
R>find.plan(PRP=c(0.05,0.95),CRP=c(0.15,0.20),
  type="hypergeom",N=500)
```

In the output below we see that the sample size should be n=51 and the acceptance number $c = 5$.

```
$n
[1] 51
$c
[1] 5
$r
[1] 6
```

The following code produces the OC curve for this plan that is shown in Figure 2.3.

```
R>library(AcceptanceSampling)
R>plan<-OC2c(51,5,type="hypergeom", N=500, pd=seq(0,.25,.01))
R>plot(plan, type='l')
R>grid()
```

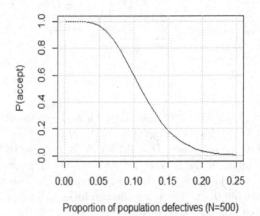

FIGURE 2.3: Operating Characteristic Curve for the plan with $N = 500$, $n = 51$, and $c = 5$

The code below shows how to determine how close the producer and consumer risk points for this plan are to the requirement.

```
R>library(AcceptanceSampling)
R>assess(OC2c(51,5), PRP=c(0.05, 0.95), CRP=c(0.15,0.20))
```

The output shows the actual probability of acceptance at AQL is 0.9589 and the probability of acceptance at RQL is 0.2032.

```
Acceptance Sampling Plan (binomial)
              Sample 1
Sample size(s)         51
Acc. Number(s)          5
Rej. Number(s)          6

Plan CANNOT meet desired risk point(s):
            Quality   RP P(accept) Plan P(accept)
PRP            0.05          0.95     0.9589318
CRP            0.15          0.20     0.2032661
```

To make the OC curve steeper and closer to the customer's ideal, the required RQL can be made closer to the AQL. For example in the R code below, the RQL is reduced from 0.15 to 0.08. As a result, the `find.plan` function finds a plan with a much higher sample size $n = 226$ (nearly 50% of the lot size $N = 500$), and acceptance number $c = 15$.

```
R>library(AcceptanceSampling)
R>find.plan(PRP=c(0.05,0.95),CRP=c(0.08,0.20),
   type="hypergeom",N=500)

$n
[1] 226
$c
[1] 15
$r
[1] 16
```

The OC curve for this plan is shown in Figure 2.4, and it is steeper with a reduced operating ratio. The disadvantage to this plan over the original ($n = 51$, $c = 5$) plan is the increased sample size $n = 226$. The next section discusses double and multiple sampling plans that can produce a steep OC curve with a smaller average sample size than required by a single sample plan.

2.5 Double and Multiple Sampling Plans

The aim of double sampling plans is to reduce the average sample size and still have the same producer and customer risk points. A double sampling plan consists of n_1, c_1, and r_1 which are the sample size, acceptance, and rejection numbers for the first sample; n_2, c_2, and r_2, which are the sample size, acceptance number, and rejection number for the second sample.

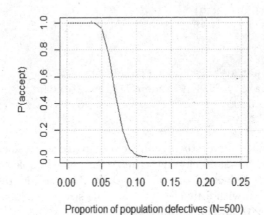

P(accept)

Proportion of population defectives (N=500)

FIGURE 2.4: Operating Characteristic Curve for the plan with $N = 500$, $n = 226$, $c = 15$

A double sampling plan consists of taking a first sample of size n_1. If there are c_1 or less nonconforming in the sample, the lot is accepted. If there are r_1 nonconforming or more in the sample, the lot is rejected (where $r_1 \geq c_1 + 2$). If the number nonconforming in the first sample is between $c_1 + 1$ and $r_1 - 1$, a second sample of size n_2 is taken. If the sum of the number of nonconforming in the first and second samples is less than or equal to c_2, the lot is accepted. Otherwise, the lot is rejected. Although there is no function in the `AcceptanceSampling` package in R for finding double sampling plans, the `assess()` function and the `OC2c()` function can be used to evaluate a double sampling plan, and the `AQLSchemes` package can retrieve double sampling plans from the ANSI/ASQ Z1.4 Standard.

Consider the following example shown by Schilling and Neubauer[84]. If a single sampling plan that has $n = 134$, and $c = 3$ is used for a lot of $N = 1000$, it will have a steep OC curve with a low operating ratio. The R code below shows that there is at least a 96% chance of accepting a lot with 1% or less nonconforming, and less than an 8% chance of accepting a lot with 5% or more nonconforming.

```
R>library(AcceptanceSampling)
R>plns<-OC2c(n=134,c=3,type="hypergeom", N=1000,
    pd=seq(0,.20,.01))
R>assess(plns,PRP=c(.01,.95),CRP=c(.05,.10))
```

```
Acceptance Sampling Plan (hypergeom)

                Sample 1
Sample size(s)      134
Acc. Number(s)        3
Rej. Number(s)        4


Plan CAN meet desired risk point(s):

            Quality   RP P(accept) Plan P(accept)
PRP           0.01          0.95      0.96615674
CRP           0.05          0.10      0.07785287
```

However, the sample size ($n=134$) is large, over 13% of the lot size. If a double sampling plan with $n_1 = 88$, $c_1 = 1$, $r_1 = 4$, and $n_2 = 88$, $c_2 = 4$, $r_2 = 5$ is used instead, virtually the same customer risk will result, and slightly less risk for the producer. This is illustrated by the R code below.

```
R>library(AcceptanceSampling)
R>pln3<-OC2c(n=c(88,88),c=c(1,4),r=c(4,5),type="hypergeom",
  N=1000,pd=seq(0,.20,.01))
R>assess(pln3,PRP=c(.01,.95),CRP=c(.05,.10))
Acceptance Sampling Plan (hypergeom)


                Sample 1 Sample 2
Sample size(s)        88       88
Acc. Number(s)         1        4
Rej. Number(s)         4        5


Plan CAN meet desired risk point(s):

            Quality   RP P(accept) Plan P(accept)
PRP           0.01          0.95       0.9805612
CRP           0.05          0.10       0.0776524
```

In the code above, the first argument to the OC2c() function, n=c(88,88) specifies n_1 and n_2 for the double sampling plan. The second argument c=c(1,4) specifies c_1 and c_2, and the third argument r=c(4,5) specifies r_1 and r_2. Notice that $r_2 = c_2 + 1$ because a decision must be made after the second sample.

The sample size for a double sampling plan will vary between n_1 and n_1+n_2 depending on whether the lot is accepted or rejected after the first sample. The probability of accepting or rejecting after the first sample depends upon the number of nonconforming items in the lot, therefore the average sample number (ASN) for the double sampling plan will be:

$$ASN = n_1 + n_2 \times P(c_1 < x_1 < r_1) \tag{2.2}$$

where x_1 is the number of nonconforming items found in the first sample.

The R code below creates Figure 2.5 that compares the sample size for the single sampling plan with the average sample for a double sampling plan at various proportions nonconforming or defective in the lot.

```
R>library(AcceptanceSampling)
R>D=seq(0,200,5) #Number nonconforming in the lot of 1000
R>pd<-D/1000  #Proportion nonconforming in the lot of 1000
R>OC1<-phyper(1, m=D, n=1000-D, k=88, lower.tail=TRUE)
R>#Probability of accepting after the first sample
R> R1<-phyper(3, m=D, n=1000-D, k=88, lower.tail=FALSE)
R>#Probability of rejecting after the first sample
R>P<-OC1+R1
R>ASN=88+88*(1-P)
R>plot(pd,ASN,type='l',ylim=c(5,150),xlab="Proportion
      nonconforming in the lot")
R>abline(134,0,lty=2)
R>text(.10,142,'single sample n=134')
R>text(.10,70,'double sample n=(88,88)')
R>grid()
```

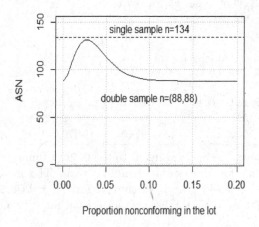

FIGURE 2.5: Comparison of Sample Sizes for Single and Double Sampling Plan

From this figure it can be seen that the average combined sample size for the double sampling plan is uniformly less than the equivalent single sampling plan. The average sample size for the double sampling plan saves most when the proportion nonconforming in the lot is less than the AQL or greater than the RQL.

The disadvantage of a double sampling plan is that they are very difficult or impossible to apply when the testing or inspection takes a long time or must be performed off site. For example, food safety and microbiological tests may take 2 to 3 days for obtaining the result.

Multiple sampling plans extend the logic of double sampling plans, by further reducing the average sample size. Multiple sampling plans can be presented in tabular form as shown in Table 2.1

If $x_1 \leq c_1$ (where x_1 is the number of nonconforming items found in the first sample) the lot is accepted. If $x_1 \geq r_1$ the lot is rejected, and if $c_1 < x_1 < r_1$ another sample is taken, etc.

A multiple sampling plan with a similar OC curve as a double sampling plan will have an even lower ASN curve than the double sampling plan. The multiple sampling plan shown in Table 2.2, has an OC curve that is very similar to the single ($n=134$, $c=3$) and the double sampling plan ($n_1=88$, $n_2=88$, $c_1=1$, $c_2=4$, $r_1=4$, $r_2=5$) presented above. Figure 2.6 shows a comparison of their OC curves magnifying on the region where these curves are steepest. It can be seen that within the region of the AQL=0.01 and the RQL $= 0.05$, these OC curves are very similar. The ASN curve for the multiple sampling plan can be shown to fall below the ASN curve for the double sampling plan shown in Figure 2.5.

Although the `AcceptanceSampling` package does not have a function for creating double or multiple sampling plans for attributes, the ANSI/ASQ Z1.4 tables discussed in Section 2.7.2 present single, double, and multiple sampling plans with matched OC curves. The single, double, and multiple sampling plans in these tables can be accessed with the `AQLSchemes` package or with the [sqc online calculator](https://www.sqconline.com/). The tables also present the OC curves and ASN curves for these plans, but the same the OC and ASN curves can be obtained from the `AQLSchemes` package as well.

TABLE 2.1: A Multiple Sampling Plan

Sample	Samp. Size	Cum. Sample Size	Accept. Number	Reject. Number
1	n_1	n_1	c_1	r_1
2	n_2	$n_1 + n_2$	c_2	r_2
$\vdots$	$\vdots$	$\vdots$	$\vdots$	$\vdots$
k	n_k	$n_1 + n_2 + \ldots + n_k$	c_k	$r_k = c_k + 1$

TABLE 2.2: A Multiple Sampling Plan with k=6

Sample	Samp. Size	Cum. Samp. Size	Accept. Number	Reject. Number
1	46	46	0	3
2	46	92	1	3
3	46	138	2	4
4	46	184	3	5
5	46	230	4	6
6	46	276	6	7

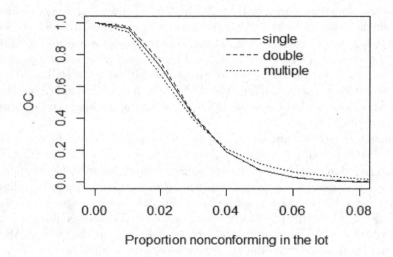

FIGURE 2.6: Comparison of OC Curves

The OC2c(), and assess() functions in the AcceptanceSampling package, and the plot() function in R can be used to evaluate the properties of a multiple sampling plan as shown in the R code below.

```
R>library(AcceptanceSampling)
R>pln4<-OC2c(n=c(46,46,46,46,46,46),c=c(0,1,2,3,4,6),
  r=c(3,3,4,5,6,7),
  type="hypergeom",N=1000,pd=seq(0,.20,.01))
R>assess(pln4,PRP=c(.01,.95),CRP=c(.05,.10))
```

```
R>plot(pln4,type='l')
R>summary(pln4,full=TRUE)
```

In summary, a sampling plan with a steeper OC curve is preferable. It benefits the customer by reducing the chances of accepting lots containing a proportion nonconforming higher than the AQL, and it motivates the supplier to send lots with a proportion nonconforming less than the AQL to avoid having lots returned. It also benefits a supplier of good quality because the probability of having a lot rejected that contains a proportion nonconforming less than the AQL is reduced.

Single sampling plans are the easiest to administer; however, the disadvantage of a single sampling plan with a steep OC curve is the additional cost of increased sampling. An equivalently steep OC curve can result from a double or multiple sampling plan with a lower average sample number (ASN). This is the advantage of double and multiple sampling plans. The disadvantage of double or multiple sampling plans is the increased bookkeeping and additional decisions that must be made.

2.6 Rectification Sampling

When rectification sampling is used, every nonconforming item found in the sample is replaced with a conforming item. Further, if the lot is rejected, the remaining items in the lot are also inspected, and any additional nonconforming items found are replaced by conforming items. In that way every rejected lot will be 100% inspected and all nonconforming items replaced with conforming items. This results in two different sampling plans (sample inspection or 100% inspection). Which one is used for a particular lot depends whether the lot is rejected based on the first sample. By use of rectification inspection, the average percent nonconforming in lots leaving the inspection station can be guaranteed to be below a certain level. This type of inspection is often used where component parts or records are produced and inspected in-house.

Assuming that an ongoing stream of lots is being inspected, the OC or probability that a lot is accepted by a single sampling plan for attributes is given by the Binomial Distribution

$$P_a = \sum_{i=0}^{c} \binom{n}{i} p^i (1-p)^{n-i} \tag{2.3}$$

where n is the sample size, c is the acceptance number and p is the probability of a nonconforming item being produced in the supplier's process. The average outgoing quality AOQ is given by

$$AOQ = \frac{P_a p (N - n)}{N} \tag{2.4}$$

where N is the lot size. The AOQ is a function of the probability of a nonconforming being produced in the supplier's process. The AOQL or average outgoing quality limit is the maximum value of AOQ. For a lot of size $N = 2000$, and a single sampling plan consisting of $n = 50$, and $c = 2$, the R code shown below, calculates the AOQ and plots it versus p. The plot is shown in Figure 2.7.

```
R>p<-seq(.01,.15,.01)
R>N<-2000
R>n<-50
R>c<-2
R>Pa<-pbinom(c,n,p,lower.tail=TRUE)
R>AOQ<-(Pa*p*(N-n))/N
R>plot(p,AOQ,type='b',xlab='Probability of nonconforming')
R>lines(c(.02,.08),c(.02639,.02639),lty=2)
R>text(.088,.02639,"AOQL")
```

From this figure, it can be seen that rectification inspection could guarantee that the average proportion nonconforming in a lot leaving the inspection station is about 0.027.

For a single sampling plan with rectification, the number of items inspected is either n or N, and the average total inspection (ATI) required is

$$ATI = n + (1 - P_a)(N - n). \tag{2.5}$$

If rectification is used with a double sampling plan,

$$ASN = n_1(P_{a_1} + P_{r_1}) + n_2(1 - P_{a_1} - P_{r_1})$$

$$\tag{2.6}$$

$$ATI = n_1 P_{a_1} + (n_1 + n_2)P_{a_2} + N(1 - P_{a_1} - P_{a_2})$$

where n_1 is sample size for the first sample, n_2 is the sample size for the second sample, P_{a_1} is the probability of accepting on the first sample, and P_{a_2} is the probability of accepting on the second sample, which is given by $\sum_{i=c_1+1}^{r_1-1} P(x_1 = i)P(x_2 \leq c_2 - x_1)$ where x_1 is the number nonconforming on the first sample and x_2 is the number nonconforming on the second sample.

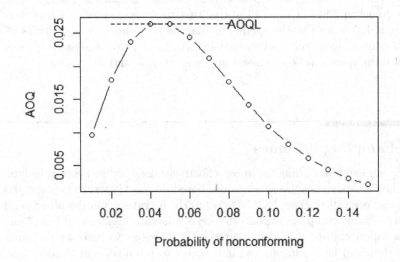

FIGURE 2.7: AOQ Curve and AOQL

2.7 Dodge-Romig Rectification Plans

Dodge and Romig developed tables for rectification plans in the late 1920s and early 1930s at the Bell Telephone Laboratories. Their tables were first published in the Bell System Technical Journal and later in book form [23]. They provided both single and double sampling for attributes.

There are two sets of plans. One set minimized the ATI for various values of the LTPD. The other set of tables minimize the ATI for a specified level of AOQL protection. The LTPD based tables are useful when you want to specify an LTPD protection on each lot inspected. The AOQL based plans are useful to guarantee the outgoing quality levels regardless of the quality coming to the inspection station. The tables for each set of plans require that the supplier's process average percent nonconforming be known. If this is not known, it can be entered as the highest level shown in the table to get a conservative plan.

The LTPD based plans provide more protection on individual lots and therefore require higher sample sizes than the AOQL based plans. While this book does not contain the Dodge Romig tables, they are available in the online sample size calculator at https://www.sqconline.com/ (note: sqconline.com offers free access for educational purposes). It provides the single sample plans for both AOQL and LTPD protection.

When used by a supply process within the same company as the customer, a more recent and better way of insuring that the average proportion noncon- forming is low is to use the quality management techniques and statistical process control techniques advocated in MIL-STD-1916. Statistical process control techniques will be discussed in Chapters 4, 5, and 6.

2.8   Sampling Schemes

Acceptance sampling plans are most effectively used for inspecting isolated lots. The OC curves for these plans are based on the Hypergeometric distri- bution and are called *Type A OC Curves* in the literature. On the other hand, when a customer company expects to receive ongoing shipments of lots from a trusted supplier, instead of one isolated lot, it is better to base the OC curve on the Binomial Distribution, and it is better to use a scheme of acceptance sampling plans (rather than one plan) to inspect the incoming stream of lots.

When a sampling scheme is utilized, there is more than one sampling plan and switching rules to dictate which sampling plan should be used at a particular time. The switching rules are based on previous samples. In a sense, the Dodge-Romig Rectification plans represent a scheme. Either a sampling inspection plan or 100% inspection is used, based on results of the sample. A sampling scheme where the switching rules are based on the result of sampling the previous lot will be described in the next section.

When basing the OC curve on the Binomial Distribution, the OC or prob- ability of acceptance is given by:

$$Pr(accept) = \sum_{i=0}^{c} \binom{n}{i} p^i (1-p)^{n-i}, \tag{2.7}$$

where n is the sample size taken from each lot, c is the acceptance number, and p is the probability of a nonconforming item being produced in the suppliers process. In the literature an OC curve based on the Binomial Distribution is called a *Type B OC Curve*.

2.9 Quick Switching Scheme

Romboski[80] proposed a straight forward sampling scheme called the quick switching scheme QSS-1. His plan proceeds as follows. There are two acceptance sampling plans. One is called the normal inspection plan consisting of sample size n and acceptance number c_N, the second is called the tightened inspection plan consisting of the same sample size n, but a reduced acceptance number c_T. The following switching rules are used:

1. Start using the normal inspection plan.

2. Switch to the tightened inspection plan immediately following a rejected lot.

3. When on tightened inspection, switch back to normal inspection immediately following an accepted lot.

4. Alternate back and forth based on these rules.

This plan can be diagrammed simply as shown in Figure 2.8.

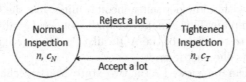

FIGURE 2.8: Romboski's QSS-1 Quick Switching System

For the case where $n = 20$, $c_N = 1$, and $c_T = 0$, Figure 2.9 compares the OC curves for the normal and tightened plans. Using the normal plan will benefit the supplier with at least a 0.94 probability of accepting lots with an average of 2% nonconforming or less. However, this plan is less than ideal for the customer who will have less than a 0.50 probability of rejecting a lot with 8% nonconforming. Using the tightened inspection plan, the customer is better protected with a more than 0.80 probability of rejecting a lot with 8% nonconforming. Nevertheless, the tightened plan OC curve is very steep in the acceptance quality range, and there is greater than a 0.32 probability of rejecting a lot with only 2% nonconforming. This would be unacceptable to most suppliers.

Using the switching rules, the quick switching scheme results in a compromise between the normal inspection plan that favors the supplier and the tightened inspection plan that benefits the customer. The OC curve for the

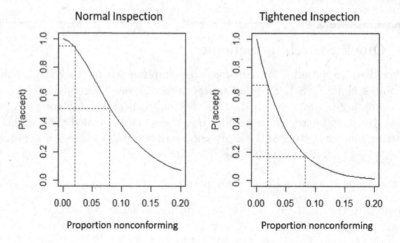

FIGURE 2.9: Comparison of Normal and Tightened Plan OC Curves

scheme will be closer to the ideal OC curve shown in Figure 2.2 without increasing the sample size.

The scheme OC curve will have a shoulder where there is a high probability of accepting lots with a low percent nonconforming, like the OC curve for the normal inspection plan. This benefits a supplier of good quality. In addition, it will drop steeply to the right of the AQL, like the OC curve for the tightened plan. This benefits the customer.

The scheme can be viewed as a two state Markov chain with the two states being normal inspection and tightened inspection. Based on this fact, Romboski[80] determined that the OC or probability of acceptance of a lot by the scheme (or combination of the two plans) was given by

$$Pr(accept) = \frac{P_T}{(1 - P_N) + P_T} \tag{2.8}$$

where P_N is the probability of accepting under normal inspection, that is given by the Binomial Probability Distribution as:

$$P_N = \sum_{i=0}^{c_N} \binom{n}{i} p^i(1-p)^{n-i}, \tag{2.9}$$

and p is the probability of a nonconforming item being produced in the suppliers process. P_T is the probability of accepting under tightened inspection, and is given by the Binomial Probability Distribution as

$$P_T = \sum_{i=0}^{c_T} \binom{n}{i} p^i (1-p)^{n-i}. \tag{2.10}$$

For the quick switching scheme with $n = 20$, $c_N = 1$, and $c_T = 0$, the R code below uses Equations (2.8) to (2.10) to calculate the OC curves for the scheme, normal, and tightened inspection, and then creates a graph shown in Figure 2.10 to compare them.

```
R># Comparison of Normal, Tightened, and QSS-1 OC Curves
R>pd<-seq(0,.20,.005)
R>PN=pbinom(1,20,pd)
R>PT=pbinom(0,20,pd)
R>Pa<-PT/((1-PN)+PT)
R>plot(pd,Pa,type='l',lty=1,xlim=c(0,.2),
     xlab='Probability of a nonconforming',ylab="OC")
R>lines(pd,PN,type='l',lty=2)
R>lines(pd,PT,type='l',lty=3)
R>lines(c(.10,.125),c(.9,.9),lty=1)
R>lines(c(.10,.125),c(.8,.8),lty=2)
R>lines(c(.10,.125),c(.7,.7),lty=3)
R>text(.15,.9,'Scheme')
R>text(.15,.8,'Normal')
R>text(.15,.7,'Tightened')
```

In this figure it can be seen that the OC curve for the QSS-1 scheme is a compromise between the OC curves for the normal and tightened sampling plans, but the sample size, $n = 20$, is the same for the scheme as it is for either of the two sampling plans.

In addition to the QSS-1 scheme resulting in an OC curve that is closer to the ideal curve shown in Figure 2.2 without increasing the sample size n, there is one additional benefit. If a supplier consistently supplies lots where the proportion nonconforming is greater than the level that has a high probability of acceptance, there is a increased chance that the sampling will be conducted with the tightened plan, and this will result in a lower probability of acceptance. If all rejected lots are returned to the supplier, this will be costly for the supplier and will be a motivation for the supplier to reduce the proportion of nonconforming. Therefore, although small customers may not be able to require procedural requirements for their suppliers to improve quality levels (as Ford Motor Company did in the early 1980's), the use of a sampling scheme to inspect incoming lots will motivate suppliers to improve quality levels themselves.

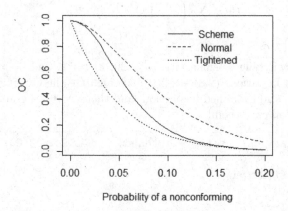

FIGURE 2.10: Comparison of Normal, Tightened, and Scheme OC Curves

2.10   MIL-STD-105E and Derivatives

The United States Military developed sampling inspection schemes as part of the World War II effort. They needed a system that did not require 100% inspection of munitions. In subsequent years improvements led to MIL-STD-105A, B, ..., E. This was a very popular system worldwide, and it was used for government and non-government contracts. The schemes in MIL-STD were based on the AQL and lot size and are intended to be applied to a stream of lots. They include sampling plans for normal, tightened, reduced sampling, and associated switching rules. Single sampling plans, double sampling plans and multiple sampling plans with equivalent OC curves are available for each lot size–AQL combination.

However, in 1995 the Army discontinued supporting the MIL-STD-105E. Civilian standards-writing organizations such as the American Standards Institute (ANSI), the International Standards Organization (ISO) and others have developed their own derivatives of the MIL-STD-105E system. The ANSI/ASQ Standard Z1.4 is the American national standard derived from MIL-STD-105E. It incorporates minor changes, none of which involved the central tables. It is recommended by the Department of Defense as a replacement and is best used for domestic contracts and in house-use. The complete document is available for purchase online at https://asq.org/quality-press/display-item?item=T964E. The R package **AQLSchemes** as well as the [sqc online calculator](https://www.sqconline.com/) contains functions that will retrieve the normal, tightened, or reduced single and double sampling plans from the AN-

SI/ASQ Z1.4 Standard. The rules for switching between Normal, tightened, and reduced sampling using ANSI/ASQ Z1.4 standard are shown in Figure 2.11. Utilizing the plans and switching rules will result in an OC curve closer to the ideal, and will motivate the suppliers to provide lots with the proportion nonconforming at or below the agreed upon AQL.

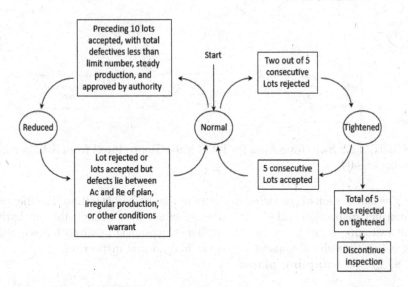

FIGURE 2.11: Switching rules for ANSI/ASQ Z1.4 (Reproduced from Schilling and Neubauer[84])

The international standard, ISO 2859-1, incorporates modifications to the original MIL-STD-105 concepts to reflect changes in the state of the art, and it is recommended for use in international trade. Again, proper use of the plans requires adherence to the switching rules which are shown in Figure 2.12. When this is done, the producer receives protection against having lots rejected when the percent nonconforming is less than the stated AQL. The customer is also protected against accepting lots with a percent nonconforming higher than the AQL. When the rules are not followed, these benefits are lost. The benefit of smaller sample sizes afforded to the customer by the reduced plan, when quality is good, is also lost when the switching rules are not followed.

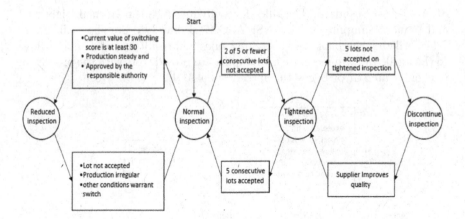

FIGURE 2.12: Switching rules for ISO 2859-1(Reproduced from Schilling and Neubauer[84])

The calculation of the switching score in Figure 2.12 is initiated at the start of normal inspection unless otherwise specified by a responsible authority. The switching score is set to zero at the start and is updated following the inspection of each subsequent lot on original normal inspection.

a) Single sampling plans:

1. When the acceptance number is 2 or more, add 3 to the switching score if the lot would have been accepted if the AQL had been one step higher; otherwise reset the switching score to zero.

2. When the acceptance number is 0 or 1, add 2 to the switching score if the lot is accepted; otherwise reset the switching score to zero.

b) Double or multiple sampling plans:

1. When a double sampling plan is used, add 3 to the switching score if the lot is accepted after the first sample; otherwise reset the switching score to 0.

2. when a multiple sampling plan is used, add 3 to the switching score if the lot is accepted by the third sample; otherwise reset the switching score to zero.

To use the tables from ANSI/ASQ Z1.4 or ISO 2859-1, a code letter is determined from a table, based on the lot size and inspection level. Next, a decision is made whether to use single, double, or multiple sampling, and whether to use normal, tightened, or reduced inspection. Finally, the sample size(s) and acceptance-rejection number(s) are obtained from the tables. The online NIST Engineering Statistics Handbook (*NIST/SEMATECH e-Handbook of Statistical Methods* - http://www.itl.nist.gov/div898/handbook/

Section 6.2.3.1) shows an example of how this is done using the MIL-STD 105E-ANSI/ASQ Z1.4 tables for normal inspection.

The section of R code below shows how a single sampling plan from AN-SI/ASQ Z1.4 or ISO 2859-1 can be retrieved using the `AASingle()` function in the R package `AQLSchemes`. The function call `AASingle('Normal')` shown below specifies that the normal sampling plan is desired. When this function call is executed, the function interactively queries the user to determine the inspection level, the lot size, and the AQL. In this example the inspection level is II, the lot size is 1500, and the AQL is 1.5%. The result is a sampling plan with $n=125$, with an acceptance number $c=5$, and a rejection number $r=6$ as shown in the section of code below.

```
R>library(AQLSchemes)
R>planS<-AASingle('Normal')
MIL-STD-105E ANSI/ASQ Z1.4
If the sample size exceeds the lot size carry out
100% inspection

What is the Inspection Level?

1: S-1
2: S-2
3: S-3
4: S-4
5: I
6: II
7: III

Selection: 6

What is the Lot Size?

 1: 2-8                2: 9-15
 3: 16-25             4: 26-50
 5: 51-90             6: 91-150
 7: 151-280           8: 281-500
 9: 501-1200         10: 1201-3200
11: 3201-10,000      12: 10,001-35,000
13: 35,001-150,000   14: 150,001-500,000
15: 500,001 and over

Selection: 10

What is the AQL in percent nonconforming per 100 items?
```

```
 1: 0.010     2: 0.015     3: 0.025     4: 0.040
 5: 0.065     6: 0.10      7: 0.15      8: 0.25
 9: 0.40     10: 0.65     11: 1.0      12: 1.5
13: 2.5      14: 4.0      15: 6.5      16: 10
17: 15       18: 25       19: 40       20: 65
21: 100      22: 150      23: 250      24: 400
25: 650      26: 1000

Selection: 12
R>planS
    n c r
1 125 5 6
```

Executing the function call again with the option tightened (i.e., `AASingle('Tightened')` and answering the queries the same as above results in the tightened plan with $n=125$, $c=3$, and $r=4$. The reduced sampling plan is obtained by executing the function `AASingle('Reduced')` and answering the queries the same way. This results in a plan with less sampling required (i.e., $n=50$, $c=2$, and $r=5$). However, there is a gap between the acceptance number and the rejection number. Whenever the acceptance number is exceeded in a ANSI/ASQ Z1.4 plan for reduced inspection, but rejection number has not been reached (for example if the number of nonconformities in a sample of 50 were 4 in the last example) then the lot should be accepted, but normal inspection should be reinstated.

One of the modifications incorporated when developing the ISO 2859-1 international derivative of MIL-STD 105E was the elimination of gaps between the acceptance and rejection numbers. the ISO 2859-1 single sampling plan, for the lot size AQL and inspection level in the example above, are the same for the normal and tightened plans, but the reduced plan is $n=50$, $c=3$, and $r=4$ with no gap between the acceptance and rejection numbers.

To create an ANSI/ASQ Z1.4 double sampling plan for the same requirements as the example above using the **AQLSchemes** package, use the **AADouble()** function as shown in the example below.

```
R>library(AQLSchemes)
R>planD<-AADouble('Normal')
MIL-STD-105E ANSI/ASQ Z1.4

What is the Inspection Level?

1: S-1
2: S-2
3: S-3
4: S-4
5: I
```

```
6: II
7: III

Selection: 6

What is the Lot Size?

  1: 2-8                    2: 9-15
  3: 16-25                  4: 26-50
  5: 51-90                  6: 91-150
  7: 151-280                8: 281-500
  9: 501-1200              10: 1201-3200
 11: 3201-10,000           12: 10,001-35,000
 13: 35,001-150,000        14: 150,001-500,000
 15: 500,001 and over

Selection: 10

What is the AQL in percent nonconforming per 100 items?

  1: 0.010    2: 0.015    3: 0.025    4: 0.040    5: 0.065
  6: 0.10     7: 0.15     8: 0.25     9: 0.40    10: 0.65
 11: 1.0     12: 1.5     13: 2.5     14: 4.0     15: 6.5
 16: 10      17: 15      18: 25      19: 40      20: 65
 21: 100     22: 150     23: 250     24: 400     25: 650
 26: 1000

Selection: 12
R>planD
          n c r
first  80 2 5
second 80 6 7
```

As can be seen in the output above, the sample sizes for the double sampling plan are 80, and 80. Accept after the first sample of 80 if there are 2 or less nonconforming, and reject if there are 5 or more nonconforming. If the number nonconforming in the first sample of 80 is 3 or 4, take a second sample of 80 and accept if the combined total of nonconforming items in the two samples is 6 or less, otherwise reject.

The OC curves are very similar for the single and double sampling plans, but not exactly the same. The OC curves for the single sampling plan ($n=125$, $c=5$) and the double sampling plan created with the AASingle() and AADouble() functions were determined separately and are compared in Figure 2.16. This figure also compares the ASN curve for the double sampling plan to the constant sample size for the single sampling plan. This figure was created with the R-code shown below. In this code the OCASNZ4S() and OCASNZ4D()

functions in the AQLSchemes package were used to create the coordinates of the OC curves and the ASN curve for the double sampling plan. The result of these functions is a data frame with columns for the proportion defective (pd) the probability of acceptance (OC) and the average sample number (ASN). The statements OCS<-SOCASN$OC, OCS<-DOCASN$OC, and ASND<-SOCASN$ASN recall the coordinates of the OC curve for the single sampling plan, the double sampling plan and the ASN for the double sampling plan respectively.

```
R>library(AQLSchemes)
R>par(mfcol=c(1,2))
R>Pnc<-seq(0,.15,.01)
R>SOCASN<-OCASNZ4S(planS,Pnc)
R>DOCASN<-OCASNZ4D(planD,Pnc)
R>OCS<-SOCASN$OC
R>OCD<-DOCASN$OC
R>ASND<-DOCASN$ASN
R># plot OC Curves
R>plot(Pnc,OCS,type='l', xlab='Proportion
  Nonconforming', ylab='OC', lty=1)
R>lines(Pnc,OCD, type='l', lty=2,col=1)
R>legend(.04,.95,c("S","D"),lty=c(1,2),col=c(1,1))
R># ASN Curves
R>plot(Pnc,ASND,type='l',ylab='ASN',lty=2,col=1,
  ylim=c(60,160))
R>lines(Pnc,rep(125,16),lty=1)
R>par(mfcol=c(1,2))
```

This figure shows that the OC curves for normal sampling with either the single or double sampling plan for the same inspection level, lot size, and AQL, are virtually equivalent. However, the double sampling plan provides slightly greater protection for the customer at intermediate levels of the proportion nonconforming. Another advantage of the double sampling plan is that the average sample size will be uniformly less than the single sampling plan.

To create ANSI/ASQ Z1.4 double sampling plans for tightened, or reduced plans use the function call AADouble('Tightened') or AADouble('Reduced') when the AQLSchemes package is loaded. Similarly the normal tightened, or reduced multiple sampling plans for ANSI/ASQ Z1.4 can be recalled with the commands AAMultiple('Normal'), AAMultiple('Tightened'), and AAMultiple('Reduced').

The OC curve for a multiple sampling plan for the same inspection level, lot size, and AQL, will be virtually the same as the OC curves for the single and double sampling plans shown in Figure 2.16, and the ASN curve will be uniformly below the ASN curve for the double sampling plan. This will be

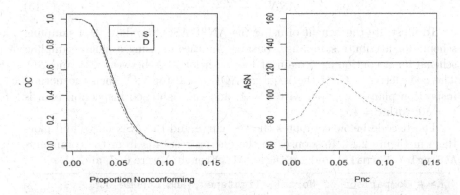

FIGURE 2.13: Comparison of OC and ASN Curves for ANSI/ASQ Z1.4 Single and Double Plans

the case for all ANSI/ASQ Z1.4 and ISO 2859-1 single double and multiple sampling plans that are matched for the same conditions.

Stephens and Larsen[92] investigated the properties of MIL-STD-105E, which is also relevant to ANSI/ASQ Standard Z1.4, and the ISO 2859-1. When ignoring the reduced plan (the use of which requires authority approval) and considering only the normal-tightend system, the sampling scheme again can be viewed as a two state Markov chain with the two states being normal inspection and tightened inspection. They showed the OC or probability of accepting by this scheme is given by the following equation:

$$Pr(accept) = \frac{aP_N + bP_T}{a + b} \qquad (2.11)$$

where P_N is the probability of accepting under normal inspection, P_T is the probability of accepting under tightened inspection, and

$$a = \frac{2 - P_N^4}{(1 - P_N)(1 - P_N^4)}$$

$$\qquad (2.12)$$

$$b = \frac{1 - P_T^5}{(1 - P_T)P_T^5}.$$

In addition, $a/(a+b)$ is the steady state probability of being in the normal sampling state, and $b/(a + b)$ is the steady state probability of being in the tightened sampling state. Therefore, the average sample number (ASN) for the normal-tightened scheme is given by the equation

$$ASN = \frac{an_N + bn_T}{a+b}.$$ (2.13)

To illustrate the benefit of using the ANSI/ASQ Standard Z1.4 sampling scheme for attribute sampling, consider the case of using a single sampling scheme for a continuing stream of lots with lot sizes between 151 and 280. The code letter is G. If the required AQL is 1.0 (or 1%), then the normal inspection plan is $n = 50$, with $c = 1$, and the tightened inspection plan is $n = 80$, with $c = 1$.

The R code below computes the OC curve and the ASN curve and plots them in Figure 2.14. Responses to the queries resulting from the commands AASingle('Normal') and AASingle('Tightened') were 6, 7, and 11.

```
R> # Comparison of Normal, Tightened, and Scheme OC
   and ASN Curves
R> planSN<-AASingle('Normal')
R> planST<-AASingle('Tightened')
R> par(mfcol=c(1,2))
R> Pnc<-seq(0,.15,.002)
R> SNOCASN<-OCASNZ4S(planSN,Pnc)
R> STOCASN<-OCASNZ4S(planST,Pnc)
R> PN<-SNOCASN$OC
R> PT<-STOCASN$OC
R> ASNN<-SNOCASN$ASN
R> ASNT<-STOCASN$ASN
R> a=(2-PN^4)/((1.0000000000001-PN)*(1.0000000000001
   -PN^4))
R> b=(1.0000000000001-PT^5)/((1.0000000000001
   -PT)*(PT^5))
R> PS<-(a*PN+b*PT)/(a+b)
R> plot(Pnc,PS,type='l',lty=1,xlim=c(0,.1),
   xlab='Probability of
   nonconforming',ylab="OC",col=1)
R> lines(Pnc,PN,type='l',lty=2,col=3)
R> lines(Pnc,PT,type='l',lty=3,col=2)
R> lines(c(.05,.06),c(.9,.9),lty=1,col=1)
R> lines(c(.05,.06),c(.8,.8),lty=2,col=3)
R> lines(c(.05,.06),c(.7,.7),lty=3,col=2)
R> text(.08,.9,'Scheme',col=1)
R> text(.08,.8,'Normal',col=3)
R> text(.08,.7,'Tightened',col=2)
R> # ASN for Scheme
R> ASN<-(a/(a+b))*ASNN+(b/(a+b))*ASNT
R> plot(Pnc,ASN,type='l',ylim=c(40,90),
   xlim=c(0,.025),xlab="Probability of nonconforming",
```

```
     ylab="ASN")
 R> lines(Pnc,ASNN,type='l',lty=2,col=3)
 R> lines(Pnc,ASNT,type='l',lty=3,col=2)
 R> par(mfcol=c(1,1))
```

The left side of Figure 2.17 shows a comparison of the OC curves for this single sampling scheme that uses the switching rules with the individual, normal, and tightened plans. Again, the OC curve for the scheme is a compromise between the normal and tightened plan. The right side of the figure shows the ASN curve for the scheme.

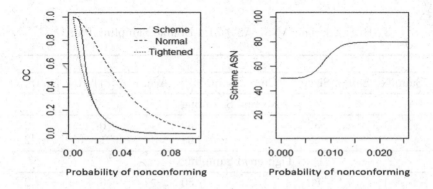

FIGURE 2.14: Comparison of Normal, Tightened, and Scheme OC Curves and ASN for Scheme

It can be seen from the figure that the OC curve for the scheme is very steep with an AQL (with 0.95 probability of acceptance) of about 0.005 or .5%, and an LTPD (with 0.10 probability of acceptance) of about 0.0285 or 2.85%. Notice the AQL is actually less than the required AQL of 1% used to find the sample sizes and acceptance numbers in the MIL-STD-105E or ANSI/ASQ Standard Z1.4 Tables. Therefore, whenever you find a normal, tightend, or reduced plan in the tables, you should check the OC curve (or a summary of the plan should be printed) to find the actual AQL and LTPD. To find a custom single sampling plan using the `find.plan()` function in the `AcceptanceSampling` package with an equivalent OC curve, with the producer risk point of (.005, .95) and customer risk point of (.0285, .10), would require a sample size $n = 233$.

It can be seen from the right side of the Figure 2.17 that the ASN for the scheme falls between $n_N = 50$ and $n_T = 80$. As the probability of a nonconforming in the lot increases beyond the LTPD, the scheme will remain in the tightened sampling plan with the sample size of 80. This is still less than half the sample size required by an equivalent single sampling plan.

For the same lot size and required AQL, the MIL-STD-105E tables also provide double and multiple sampling schemes with equivalent OC curves. The double sampling scheme for this example is shown in Table 2.3 and uses the same switching rules shown in Figure 2.11. The multiple sampling plan is shown in Table 2.4, and again uses the same switching rules.

The double sampling scheme will have a uniformly smaller ASN than the single sampling scheme, shown in Figure 2.14, and the multiple sampling scheme will have a uniformly smaller ASN than the double sampling scheme. The double and multiple sampling schemes will require more bookkeeping to administer, but they will result in reduced sampling with the same protection for producer and supplier.

TABLE 2.3: The ANSI/ASQ Z1.4 Double Sampling Plan

Sample	Samp. Size	Cum. Samp. Size	Acc. No. (c)	Rej. No.(r)
Normal Sampling				
1	32	32	0	2
2	32	64	1	2
Tightened Sampling				
1	50	50	0	2
2	50	100	1	2

In summary, when a continuing stream of lots is to be inspected from a supplier, the tabled sampling schemes can produce an OC curve closer to the ideal shown in Figure 2.2 with much reduced sampling effort.

2.11   MIL-STD-1916

In 1996, The U.S. Department of Defense issued MIL-STD-1916. It is not derived from MIL-STD-105E, and is totally unique. This standard includes attribute and variable sampling schemes in addition to guidelines on quality management procedures and quality control charts. The attribute sampling schemes in MIL-STD-1916 are zero nonconforming plans where the acceptance number c is always zero. The rational is that companies (including the Department of Defense) that use total quality management (TQM) and continuous process improvement, would not want to use AQL-based acceptance

TABLE 2.4: The ANSI/ASQ Z1.4 Multiple Sampling Plan

Sample	Samp. Size	Cum. Samp. Size	Acc. No. (c)	Rej. No.(r)
		Normal Sampling		
1	13	13	#	2
2	13	26	#	2
3	13	39	0	2
4	13	52	0	3
5	13	65	1	3
6	13	78	1	3
7	13	91	2	3
		Tightened Sampling		
1	20	20	#	2
2	20	40	#	2
3	20	60	0	2
4	20	80	0	3
5	20	100	1	3
6	20	120	1	3
7	20	140	2	3

sampling plans that allow for a nonzero level of nonconformities. The MIL-STD-1916 zero nonconforming plans, and their associated switching rules can be accessed through the online calculators at https://www.sqconline.com/.

2.12 Summary

This chapter has shown how to obtain single sampling plans for attributes to match a desired producer and consumer risk point. Double sampling plans and multiple sampling plans can have the same QC curve as a single sampling plan with a reduced sample size. The example in illustrated in Figure 2.5 shows that the average sample number for a double sampling plan was lower than the equivalent single sample plan, although it depends upon the fraction nonconforming in the lot. A multiple sampling plan will have a lower average sample number than the double sampling plan with an equivalent OC curve.

Sampling schemes that switch between one or two different plans, depending on the result of previous lots, are more appropriate for sampling a stream of lots from a supplier. Single, double, or multiple sampling schemes will result in a lower average sample number than using an OC-equivalent single, double, or multiple sampling plan for all lots in the incoming stream. Published tables of sampling schemes (like ANSI/ASQ-Z1.4 and ISO 2859-1) are recommended for in-house or domestic or international trade when the producer and consumer can agree on an acceptable quality level (AQL). In that case, use of the tabled scheme will result in the minimum inspection needed to insure the producer and consumer risks are low.

2.13   Exercises

1. Run all the code examples in the chapter, as you read.

2. Use the `find.plan()` function in the `AcceptanceSampling` package to find a single sampling plan for the situation where AQL=0.04 with $\alpha = 0.05$, RQL=0.08, and $\beta = 0.10$ for:

   (a) A lot size of 5000

   (b) A lot size of 80

3. For the plan you found for Exercise 2. (a), show the Type B OC curve and the Type A OC curve are nearly equivalent.

4. Use the `AASingle()` function in the `AQLSchemes` package to verify the normal and tightened inspection plans mentioned below equation 2.13 are the correct ANSI/ASQ Z1.4 plans.

5. If the lot size is $N = 2000$, and the AQL $= 0.65$ percent, use the `AASingle()` and `AADouble()` functions in the `AQLSchemes` package to find the ANSI/ASQ-Z1.4 single and double sampling plans for:

   (a) Normal inspection

   (b) Tightened inspection

6. Plot the OC curves for the single sampling plans you found in Exercise 5. (a) and (b).

7. Find the OC curve for the ANSI/ASQ-Z1.4 single sampling scheme (consisting of normal and tightened inspection) that you found in Exercise 5.

8. Find a single sampling plan that has an OC curve equivalent to the scheme OC curve you found in Exercise 7.

3

Variables Sampling Plans

When actual quantitative information can be measured on sampled items, rather than simply classifying them as conforming or nonconforming, variables sampling plans can be used. To achieve the same operating characteristic, a variables sampling plan requires fewer samples than an attribute plan since more information is available in the measurements. If the measurements do not require much more time and expense than classifying items, the variables sampling plan provides an advantage. When a lot is rejected, the measurements in relation to the specification limits give additional information to the supplier and may help to prevent rejected lots in the future. This is also an advantage of the variables sampling plan.

The disadvantage of variables sampling plans is that they are based on the assumption that the measurements are normally distributed (at least for the plans available in published tables or through pre-written software).

Another critical assumption of variable plans is that the measurement error, in the quantitative information obtained from sampled items, is small relative to the specification limits. If not, the type I and type II errors for the lot acceptance criterion will both be inflated. The standard way of estimating measurement error and comparing it to the specification limits is through the use of a Gauge Repeatability and Reproducibility Study or Gauge R&R study. Section 3.4 will illustrate a basic Gauge R&R study and show R code to analyze the data.

To illustrate the idea of a variables sampling plan, consider the following example. A granulated product is delivered to the customer in lots of 5000 bags. A lower specification limit on the particle size is $LSL=10$. A variables sampling plan consists of taking a sample of bags and putting the granulated material in the sampled bags through a series of meshes in order to determine the smallest particle size. This is the measurement. An attribute sampling plan would simply classify each sampled bag containing any particles of size less than the $LSL=10$ to be nonconforming or defective, and any bag whose smallest particles are greater than the LSL to be conforming.

There are two different methods for developing the acceptance criteria for a variables sampling plan. The first method is called the k-Method. It compares the standardized difference between the specification limit and the mean of the measurements made on each sampled item to an acceptance constant k. The second method is called the M-Method. It compares the estimated proportion of items out of specifications in the lot to a maximum allowable proportion M. When there is only one specification limit (i.e., USL or LSL) the k-method and the M-method yield the same results. When there is both an upper and lower specification limit, the M-method must be used in all but special circumstances.

3.1   The k-Method

3.1.1   Lower Specification Limit

3.1.1.1   Standard Deviation Known

To define a variables sampling plan the number of samples (n) and an acceptance constant (k) must be determined. A lot would be accepted if $(\bar{x} - LSL)/\sigma > k$, where $\bar{x}$ is the sample average of the measurements from a sample and σ is the standard deviation of the measurements. The mean and standard deviation are assumed to be known from past experience. In terms of the statistical theory of hypothesis testing, accepting the lot would be equivalent to failing to reject the null hypothesis $H_0 : \mu \geq \mu_{AQL}$ in favor of the alternative $H_a : \mu < \mu_{AQL}$.

When the measurements are assumed to be normally distributed with a lower specification limit LSL, the AQL and the RQL in terms of proportion of items below the LSL can be visualized as areas under the normal curve to the left of the LSL as shown in Figure 3.1. In this figure it can be seen that when the mean is μ_{AQL} the proportion of defective items is AQL, and when the mean of the distribution is μ_{RQL} the proportion of defective items is RQL.

If the producer's risk is α and the consumer's risk is β, then

$$P\left(\frac{\bar{x} - LSL}{\sigma} > k \mid \mu = \mu_{AQL}\right) = 1 - \alpha \tag{3.1}$$

and

$$P\left(\frac{\bar{x} - LSL}{\sigma} > k \mid \mu = \mu_{RQL}\right) = \beta. \tag{3.2}$$

In terms statistical hypothesis testing, Equation 3.1 is equivalent to having a significance level of α for testing $H_0 : \mu \geq \mu_{AQL}$, and Equation 3.2 is equivalent to having power of $1 - \beta$ when $\mu = \mu_{RQL}$.

By multiplying both sides of the inequality inside the parenthesis in Equation 3.1 by $\sqrt{n}$, subtracting $\frac{\mu_{AQL}}{\sigma/\sqrt{n}}$ from each side, and adding $\frac{LSL}{\sigma/\sqrt{n}}$ to each side, it can be seen that

$$\frac{\bar{x} - LSL}{\sigma} > k \Rightarrow \frac{\bar{x} - \mu_{AQL}}{\sigma/\sqrt{n}} > k\sqrt{n} + \frac{LSL - \mu_{AQL}}{\sigma/\sqrt{n}}. \tag{3.3}$$

If the mean is given as $\mu = \mu_{AQL}$, then $\frac{\bar{x} - \mu_{AQL}}{\sigma/\sqrt{n}}$ follows the standard normal

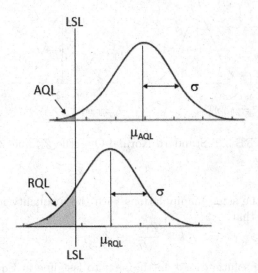

FIGURE 3.1: AQL and RQL for Variable Plan

distribution with mean $\mu = 0$ and standard deviation $\sigma = 1$. Therefore,

$$P\left(Z > k\sqrt{n} + \frac{LSL - \mu_{AQL}}{\sigma/\sqrt{n}}\right) = 1 - \alpha,$$

and   (3.4)

$$P\left(Z < k\sqrt{n} + \frac{LSL - \mu_{AQL}}{\sigma/\sqrt{n}}\right) = \alpha.$$

Consequently,

$$k\sqrt{n} + \frac{LSL - \mu_{AQL}}{\sigma/\sqrt{n}} = Z_\alpha$$

or

(3.5)

$$k = \frac{Z_\alpha}{\sqrt{n}} - \frac{LSL - \mu_{AQL}}{\sigma} = \frac{Z_\alpha}{\sqrt{n}} - Z_{AQL},$$

since $\dfrac{LSL - \mu_{AQL}}{\sigma} = Z_{AQL}.$

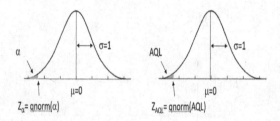

FIGURE 3.2: Standard Normal Quantile Z_α and Z_{AQL}

Performing the same manipulations with the inequality in Equation 3.2, it can be shown that

$$k = \frac{Z_{1-\beta}}{\sqrt{n}} - Z_{RQL}.$$
(3.6)

Equating the solution for k in the next to last line in Equation 3.5 with the solution for k in Equation 3.6 and solving for n, it can be seen that

$$\frac{Z_\alpha}{\sqrt{n}} - Z_{AQL} = \frac{Z_{1-\beta}}{\sqrt{n}} - Z_{RQL} \Rightarrow n = \left(\frac{Z_\alpha - Z_{1-\beta}}{Z_{AQL} - Z_{RQL}}\right)^2.$$
(3.7)

Z_α, Z_{AQL}, and Z_{RQL} are quantiles of the standard normal distribution, as shown in Figure 3.2. They can be calculated with the R function qnorm. For the case where AQL=1% or 0.01, the RQL=4.6% or 0.046, $\alpha = 0.05$ and $\beta = 0.10$, the R Code below evaluates Equation 3.7 and the last line of Equation 3.5 to find the sample size n and the acceptance constant k for a custom variable sampling plan that has PRP=(.01,.95), and CRP=(.046,.10).

```
R>n<-((qnorm(.05)-qnorm(.90))/(qnorm(.01)-qnorm(.046)))^2
R>k<-(qnorm(.05) / sqrt(21))-qnorm(.01)
```

The first line of code calculates $n = 20.8162$. This is rounded up to the next integer (21) which is substituted into the next line for for n in calculating $k = 1.967411$.

Thus, conducting the sampling plan on a lot of material consists of the following steps:

1. Take a random sample of n items from the lot

2. Measure the critical characteristic x on each sampled item

3. Calculate the mean measurement $\bar{x}$

4. Compare $(\bar{x} - LSL)/\sigma$ to the acceptance constant $k = 1.967411$

5. If $(\bar{x} - LSL)/\sigma > k$, accept the lot, otherwise reject the lot.

The R code below uses the `OCvar()` function in the `AcceptanceSampling` package to store the plan and make the OC curve shown in Figure 3.3.

```
R>library(AcceptanceSampling)
R>plnVkm<-OCvar(21,1.967411,pd=seq(0,.15,.001))
R>plot(plnVkm, type='l')
R>grid()
```

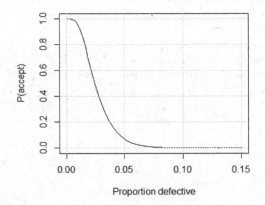

FIGURE 3.3: OC Curve for Variables Sampling Plan $n = 21$, $k = 1.967411$

This OC curve is very steep and close to the ideal with a very high probability of accepting lots with 1% defective or less, and a very low probability of accepting lots with 4.6% defective or more.

The `find.plan()` function in the `AcceptanceSampling` package automates the procedure of finding the sample size (n) and acceptance constant (k) for a custom derived variables sampling plan. The R code and output below shows how this is done.

```
R>library(AcceptanceSampling)
R>find.plan(PRP=c(0.01, 0.95),CRP=c(0.046, 0.10),
  type="normal")
R> $n
[1] 21

$k
[1] 1.967411

$s.type
```

[1] "known"

If the items in the sample could only be classified as nonconforming or conforming, an attribute sampling plan would require many more samples to have a equivalent OC curve. The R code and output below uses the `find.plan()` function in the `AcceptanceSampling` package to find an equivalent custom derived attribute sampling plan and plot the OC curve shown in Figure 3.4. The `type="binomial"` option in the `find.plan()` was used assuming that the lot size is large enough that the Binomial distribution is an accurate approximation to the Hypergeometric distribution for calculating the probability of acceptance with the attribute plan.

```
R>library(AcceptanceSampling)
R>find.plan(PRP=c(0.01,0.95),CRP=c(0.046,0.10),
  type="binomial")
$n
[1] 172

$c
[1] 4

$r
[1] 5


R>plnA<-OC2c(n=172,c=4, type="binomial",pd=seq(0,.15,.001))
R>plot(plnA,type='l', xlim=c(0,.15))
R>grid()
```

We can see that the OC curve for the attribute sampling plan in Figure 3.4 is about the same as the OC curve for the variables sampling plan in Figure 3.3, but the sample size $n = 172$ required for the attribute plan is much higher than the sample size $n = 21$ for the variable plan. The variable plan would be advantageous unless the effort and expense of making the measurements took more than 8 times the effort and expense of simply classifying the items as conforming to specifications or not.

Example 1 Mitra[70] presents the following example of the use of a variables sampling plan with a lower specification limit and the standard deviation known. In the the manufacture of heavy-duty utility bags for household use, the lower specification limit on the carrying weight is 100 kg. The AQL is 0.02 or 2%, the RQL is 0.12 or 12%. The producers risk $\alpha = 0.08$, and the consumers risk $\beta = 0.10$, and the standard deviation of the carrying weight is

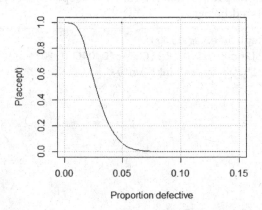

FIGURE 3.4: OC Curve for Attribute Sampling Plan $n = 172$, $c = 4$

$\sigma = 8$ kg. The R Code and output below shows the sample size $n = 10$, and the acceptance constant $k = 1.6009$ for the custom derived sampling plan.

```
R>library(AcceptanceSampling)
R>find.plan(PRP=c(.02,.92),CRP=c(.12,.10),type='normal')
$n
[1] 10

$k
[1] 1.609426

$s.type
[1] "known"
```

After selecting a sample of $n = 10$ bags, the average carrying weight was found to be $\bar{x} = 110$. Then,

$$Z_L = \frac{\bar{x} - LSL}{\sigma} = \frac{110 - 100}{8} = 1.25 < 1.609 = k$$

Therefore, reject the lot.

3.1.1.2 Standard Deviation Unknown

If it could not be assumed that the standard deviation σ were known, the Z_α and $Z_{1-\beta}$ in Equation 3.7 would have to be replaced with the quantiles of the t-distribution with $n - 1$ degrees of freedom (t_α and $t_{1-\beta}$). Then Equation 3.7 could not directly be solved for n, because the right side of the equation is also a function n. Instead an iterative approach would have to be used to solve for n. The `find.plan()` function in the `AcceptanceSampling` package does this as illustrated below.

```
R>library(AcceptanceSampling)
R>find.plan(PRP=c(0.01,0.95),CRP=c(0.046,0.10),
   type="normal", s.type="unknown")
$n
[1] 63

$k
[1] 1.974026

$s.type
[1] "unknown"
```

The sample size $n = 63$ for this plan with σ unknown is still much less than the $n = 172$ that would be required for the attribute sampling plan with an equivalent OC curve.

The R code below uses the `OC2c()`, and `OCvar()` functions in the `AcceptanceSampling` package to store the plans for the attribute sampling plan (whose OC curve is shown in Figure 3.4) and the variable sampling plans for the cases where σ is unknown or known. The values operating characteristic OC can be retrieved from each plan by attaching the suffix `@paccept` to the names for the stored plans. The OC curves for each plan are then plotted on the same graph in Figure 3.5 using the R `plot` function for comparison.

```
R>library(AcceptanceSampling)
R>PD<-seq(0,.15,.001)
R>plnA<-OC2c(n=172,c=4,type="binomial",pd=PD)
R>plnVku<-OCvar(n=63,k=1.974026,pd=PD,s.type="unknown")
R>plnVkm<-OCvar(n=21,k=1.96411,pd=PD,s.type="known")
R>#Plot all three OC curves on the same graph
R>plot(PD,plnA@paccept,type='l',lty=1,col=1,
   xlab='Probability of nonconforming',ylab='OC')
R>lines(PD,plnVku@paccept,type='l',lty=2,col=2)
R>lines(PD,plnVkm@paccept,type='l',lty=4,col=4)
R>legend(.04,.95,c("Attribute","Variable, sigma  unknown",
   "Variable, sigma known"),lty=c(1,2,4),col=c(1,2,4))
R>grid()
```

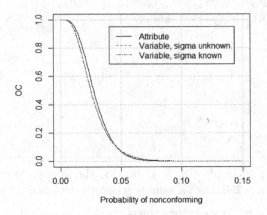

FIGURE 3.5: Comparison of OC Curves for Attribute and Variable Sampling Plans

It can be seen in Figure 3.5 that the OC curve for the two variable plans are essentially identical, although the sample size (n=21) is less when σ is known. The OC curves for these variable plans are very similar to the OC curve for the attribute plan (that is also shown in Figure 3.4), but they do offer the customer slightly more protection for intermediate values of the proportion nonconforming, as can be seen in Figure 3.5.

When the standard deviation is unknown, conducting the sampling plan on a lot of material consists of the following steps:

1. Take a random sample of n items from the lot

2. Measure the critical characteristic x on each sampled item

3. Calculate the mean measurement $\bar{x}$, and the sample standard deviation s

4. Compare $(\bar{x} - LSL)/s$ to the acceptance constant k

5. If $(\bar{x} - LSL)/s > k$, accept the lot, otherwise reject the lot.

Example 2 Montgomery[72] presents the following example of the use of a custom derived variables sampling plan with a lower specification limit and standard deviation unknown. A soft drink bottler buys nonreturnable bottles from a supplier. Their lower specification limit on the bursting strength is 225psi. The AQL is 0.01 or 1%, the RQL is 0.06 or 6%. The producers risk $\alpha = 0.05$, and the consumers risk $\beta = 0.10$, and the standard deviation of the bursting strength is unknown. The R Code and output below show the sample size $n = 42$, and the acceptance constant $k = 1.905285$.

```
R>library(AcceptanceSampling)
R>find.plan(PRP=c(.01,.95), CRP=c(.06,.10),
  type='normal',s.type='unknown')
$n
[1] 42

$k
[1] 1.905285

$s.type
[1] "unknown"
```

After selecting a sample of $n = 42$ bottles, the average bursting strength was found to be $\bar{x} = 255$, with a sample standard deviation of $s = 15$. Then,

$$Z_L = \frac{\bar{x} - LSL}{s} = \frac{255 - 225}{15} = 2.0 > 1.905285 = k$$

Therefore, accept the lot.

3.1.2   Upper Specification Limit

3.1.2.1   Standard Deviation Known

If a variables sampling plan was required for a situation where there was an upper specification limit (USL), instead of a lower specification limit (LSL), then the fourth and fifth steps in conducting the sampling plan on a lot of material would change from:

4. Compare $(\bar{x} - LSL)/\sigma$ to the acceptance constant k

5. If $(\bar{x} - LSL)/\sigma > k$, accept the lot, otherwise reject the lot.

   to

4. Compare $(USL - \bar{x})/\sigma$ to the acceptance constant k

5. If $(USL - \bar{x})/\sigma > k$, accept the lot, otherwise reject the lot.

3.1.2.2   Standard Deviation Unknown

If the standard deviation were unknown, steps 4 and 5 above would change to

4. Compare $(USL - \bar{x})/s$ to the acceptance constant k

5. If $(USL - \bar{x})/s > k$, accept the lot, otherwise reject the lot.

and the sample size n and acceptance constant k would be found with the find.plan() function, as shown in the example code below.

```
R>library(AcceptanceSampling)
R>find.plan(PRP=c(0.01, 0.95), CRP=c(0.06, 0.10),
  type="normal", s.type="unknown")
```

3.1.3 Upper and Lower Specification Limits

3.1.3.1 Standard Deviation Known

When there is an upper (USL) and a lower specification limit (LSL), Schilling[84] proposed a simple procedure that may be used to determine if two separate single specification limit plans may be used. The procedure is as follows:

1. Calculate $Z_{p*} = (LSL - USL)/2\sigma$

2. Calculate $p* = \text{pnorm}(Z_{p*})$ the area under the standard normal density to the left of Z_{p*}.

3. If $2p* \geq RQL$ reject the lot, because even if the distribution is centered between the specification limits, the a proportion outside the specification limits will be too high.

4. If $2p* \leq AQL$ use two single specification sampling plans (i.e., one for LSL and one for USL).

5. If $AQL \leq 2p* \leq RQL$ then use the M-Method for upper and lower specification limits that is described in Section 3.2

3.1.3.2 Standard Deviation Unknown

When there is an upper (USL) and a lower specification limit (LSL), and the standard deviation is unknown, use the M-Method described in Section 3.2.

3.2 The M-Method

3.2.1 Lower Specification Limit

3.2.1.1 Standard Deviation Known

For a variables sampling plan, the M-Method compares the estimated proportion below the *LSL* to a maximum allowable proportion. In this case, the uniform minimum variance unbiased estimate of the proportion below the *LSL* developed by Lieberman and Resnikoff[64] is used. The maximum allowable proportion below *LSL* is a function of the acceptance constant k used in the k-method. For the case of a lower specification limit and the standard deviation known, the uniform minimum variance unbiased estimate of the proportion defective is:

$$P_L = \int_{Q_L}^{\infty} \frac{1}{\sqrt{2\pi}} e^{-t^2/2} dt, \tag{3.8}$$

or the area under the standard normal distribution to the right of $Q_L = Z_L \left(\sqrt{\frac{n}{n-1}} \right)$,

where $Z_L = (\bar{x} - LSL)/\sigma$. The maximum allowable proportion defective is:

$$M = \int_{k\sqrt{\frac{n}{n-1}}}^{\infty} \frac{1}{\sqrt{2\pi}} e^{-t^2/2} dt, \tag{3.9}$$

or the area under the standard normal distribution to the right of $k\sqrt{\frac{n}{n-1}}$, where k is the acceptance constant used in the k-method.

Example 3 To illustrate, reconsider Example 1 from Mitra[70]. The sample size was $n = 10$, and the acceptance constant was $k = 1.6094$. The lower specification limit was $LSL = 100$, and the known standard deviation was $\sigma = 8$. In the sample of 10, $\bar{x} = 110$. Therefore, $Q_L = \left(\frac{110-100}{8} \right) \left(\sqrt{\frac{10}{9}} \right) = 1.3176$. The estimated proportion defective and the R command to evaluate it is:

$$P_L = \int_{1.3176}^{\infty} \frac{1}{\sqrt{2\pi}} e^{-t^2/2} dt = \texttt{1-pnorm(1.3176)} = 0.0938.$$

The maximum allowable proportion defective and the R command to evaluate it is:

$$M = \int_{1.6094\sqrt{\frac{10}{9}}}^{\infty} \frac{1}{\sqrt{2\pi}} e^{-t^2/2} dt = \texttt{1-pnorm(1.6094*sqrt(10/9))} = 0.0449.$$

Since $0.0938 > 0.0449$ (or $P_L > M$), reject the lot. This is the same conclusion reached with the k-method shown in Example 1. When there is only one specification limit the results of the k-method and the M-method will always agree.

M and P_L can be calculated with the R function **pnorm** or alternatively with the **MPn** and **EPn** functions in the R package **AQLSchemes** as shown in the code below.

```
R>M<-pnorm(1.6094*sqrt(10/9),lower.tail=F)
R>PL<-pnorm(((110-100)/8)*sqrt(10/9),lower.tail=F)
R>M
[1] 0.04489973
R>PL
[1] 0.09381616
library(AQLSchemes)
R>PL<-EPn(sided="one",stype="known",LSL=100,sigma=8,xbar=110,
  n=10)
```

```
R>PL
[1] 0.09381616
R>M<-MPn(k=1.6094,n=10,stype="known")
R>M
[1] 0.04489973
```

3.2.1.2 Standard Deviation Unknown

When the standard deviation is unknown, the symmetric standardized Beta distribution is used instead of the standard normal distribution in calculating the uniform minimum variance unbiased estimate of the proportion defective. The standardized Beta CDF is defined as

$$B_x(a,b) = \frac{\Gamma(a+b)}{\Gamma(a)\Gamma(b)} \int_0^x \nu^{a-1}(1-\nu)^{b-1}d\nu, \tag{3.10}$$

where $0 \leq x \leq 1$, $a > 0$, and $b > 0$. The density function is symmetric when $a = b$. The estimate of the proportion defective is then

$$\hat{p}_L = B_x(a,b), \tag{3.11}$$

where $a = b = \frac{n}{2} - 1$, and

$$x = \max\left(0, .5 - .5Z_L\left(\frac{\sqrt{n}}{n-1}\right)\right),$$

and the sample standard deviation s is substituted for σ in the formula for

$$Z_L = \frac{\bar{x} - LSL}{s}.$$

When the standard deviation is unknown, the maximum allowable proportion defective is calculated as:

$$M = B_{B_M}\left(\frac{n-2}{2}, \frac{n-2}{2}\right), \tag{3.12}$$

where

$$B_M = .5\left(1 - k\frac{\sqrt{n}}{n-1}\right),$$

and k is the acceptance constant. The Beta CDF function can be evaluated with R using the $B_x(a,b) = $ pbeta(x,a,b) function.

Example 4 To illustrate, reconsider Example 2 from Montgomery[72]. The sample size was $n = 42$, and the acceptance constant was $k = 1.905285$. The lower specification limit was $LSL = 225$. In the sample of 42, $\bar{x} = 255$, and the sample standard deviation was $s = 15$. Therefore,

$$Z_L = \left(\frac{255 - 225}{15}\right) = 2.0.$$

The estimated proportion defective and the R command to evaluate it is:

$$\hat{p}_L = B_x(a,b) = \text{pbeta}(.3419322,20,20) = 0.02069563,$$

where $a = b = \frac{42}{2} - 1 = 20$, and

$$x = \max\left(0, .5 - .5(2.0)\left(\frac{\sqrt{42}}{42-1}\right)\right) = 0.3419332.$$

The maximum allowable proportion defective and the R command to evaluate it is:

$$M = B_{B_M}\left(\frac{42-2}{2}, \frac{42-2}{2}\right) = \text{pbeta}(.3494188,20,20) = 0.02630455,$$

where

$$B_M = .5\left(1 - 1.905285\left(\frac{\sqrt{42}}{42-1}\right)\right) = 0.3494188.$$

Since $0.02069563 < 0.02630455$ (or $\hat{p}_L < M$), accept the lot. This is the same conclusion reached with the k-method shown in Example 2.

The calculation of the estimated proportion defective, $\hat{p}_L$, and the maximum allowable proportion defective, M, can be simplified using the EPn() and MPn() functions in the R package AQLSchemes as shown in the R code below.

```
R>library(AQLSchemes)
R>PL<-EPn(sided="one",stype="unknown",LSL=225,xbar=255,
  s=15,n=42)
R>PL
[1] 0.02069563
R>M<-MPn(k=1.905285,stype="unknown",n=42)
R>M
[1] 0.02630455
```

3.2.2 Upper Specification Limit

3.2.2.1 Standard Deviation Known

When there is an upper specification limit and the standard deviation is known, the acceptance criterion changes from $P_L < M$ to $P_U < M$, where

$$P_U = \int_{Q_U}^{\infty} \frac{1}{\sqrt{2\pi}} e^{-t^2/2} dt, \tag{3.13}$$

or the area under the standard normal distribution to the right of $Q_U = Z_U\left(\sqrt{\frac{n}{n-1}}\right)$ and $Z_U = (USL - \bar{x})/\sigma$.

3.2.2.2 Standard Deviation Unknown

When there is an upper specification limit and the standard deviation unknown, the acceptance criterion is $\hat{p}_U < M$ where

$$\hat{p}_U = B_x(a, b), \tag{3.14}$$

$$a = b = \frac{n}{2} - 1,$$

$$x = \max\left(0, .5 - .5Z_U\left(\frac{\sqrt{n}}{n-1}\right)\right),$$

$$Z_U = \frac{USL - \bar{x}}{s},$$

and M is the same as that defined in Equation 3.12.

3.2.3 Upper and Lower Specification Limit

3.2.3.1 Standard Deviation Known

When there are both upper and lower specification limits and the standard deviation is known, the acceptance criterion becomes: accept if

$$P = (P_L + P_U) < M, \tag{3.15}$$

where P_L is defined in Equation 3.8, P_U is defined in Equation 3.13, and M is defined in Equation 3.9.

3.2.3.2 Standard Deviation Unknown

When the standard deviation is unknown, the acceptance criterion becomes: accept if

$$\hat{p} = (\hat{p}_L + \hat{p}_U) < M, \tag{3.16}$$

where $\hat{p}_L$ is defined in Equation 3.11, $\hat{p}_U$ is defined in Equation 3.14 and M is defined in Equation 3.12.

Example 5 Reconsider the variables sampling plans whose OC curves were shown in Figure 3.3 where σ was known, and Figure 3.5 where σ was unknown. Suppose the upper specification limit was $USL = 100$, and the lower specification limit was $LSL = 90$.

When the standard deviation was known to be $\sigma = 2.0$, the sample size was $n = 21$, and the acceptance constant $k = 1.967411$, as indicated in Figure 3.3. If $\bar{x}$ was determined to be 96.68 after taking a sample of 21, then

$$Q_U = \left(\frac{(100 - 96.68)}{2.0}\right)\sqrt{\frac{21}{20}} = 1.701,$$

and

$$P_U = \int_{1.701}^{\infty} \frac{1}{\sqrt{2\pi}} e^{-t^2/2} dt = 0.04447.$$

$$Q_L = \left(\frac{(96.68 - 90)}{2.0} \right) \sqrt{\frac{21}{20}} = 3.4225,$$

and

$$P_L = \int_{3.4225}^{\infty} \frac{1}{\sqrt{2\pi}} e^{-t^2/2} dt = 0.00031.$$

$$M = \int_{1.967411\sqrt{\frac{21}{20}}}^{\infty} \frac{1}{\sqrt{2\pi}} e^{-t^2/2} dt = 0.0219.$$

Therefore, $P = (P_L + P_U) = 0.0448 > 0.0219 = M$, and the decision would be to reject the lot.

P and M can again be calculated using the EPn() and MPn() functions as shown below.

```
R>library(AQLSchemes)
R># sigma known
R>P<-EPn(sided="two",stype="known",sigma=2,LSL=90,
  USL=100,xbar=96.68,n=21)
R>P
[1] 0.04478233
R>M<-MPn(k=1.967411,stype="known",n=21)
R>M
[1] 0.02190018
```

When the standard deviation was unknown (as in Figure 3.5), the sample size was $n = 63$, and the acceptance constant $k = 1.97403$, as indicated in Figure 3.5. If $\bar{x}$ was determined to be 97.006, and the sample standard deviation was 1.9783 after taking a sample of 63, then

$$\hat{p}_U = B_x(a, b) = 0.06407, \tag{3.17}$$

where

$$a = b = \frac{63}{2} - 1 = 30.5,$$

$$x = \max \left(0, .5 - .5 Q_U \left(\frac{\sqrt{63}}{63 - 1} \right) \right) = 0.4031,$$

and

$$Q_U = \frac{100 - 97.006}{1.9783} = 1.51342.$$

$$\hat{p}_L = B_x(a, b) = 0.000095, \tag{3.18}$$

where

$$a = b = \frac{63}{2} - 1 = 30.5,$$

$$x = \max\left(0, .5 - .5Q_L\left(\frac{\sqrt{63}}{63 - 1}\right)\right) = 0.2733,$$

and

$$Q_L = \frac{97.006 - 90}{1.9783} = 3.541.$$

$$M = B_{B_M}\left(\frac{63 - 2}{2}, \frac{63 - 2}{2}\right) = 0.02284, \tag{3.19}$$

where $B_M = .5\left(1 - 1.97403\frac{\sqrt{63}}{63-1}\right) = 0.37364$.

Therefore, $\hat{p} = (\hat{p}_L + \hat{p}_U) = 0.06416 > 0.02284 = M$, and again the decision would be to reject the lot.

$\hat{p}$ and M can again be calculated using the EPn() and MPn() functions as shown below.

```
R>library(AQLSchemes)
R># sigma unknown
R>P<-EPn(sided="two",stype="unknown",LSL=90,USL=100,
   xbar=97.006,s=1.9783,n=63)
R>P
[1] 0.06416326
R>M<-MPn(k=1.97403,stype="unknown",n=63)
R>M
[1] 0.02284391
```

3.3 Sampling Schemes

3.3.1 MIL-STD-414 and Derivatives

Based on the work of Lieberman and Resnikoff[64], the U.S. Department of Defense issued the AQL based MIL-STD-414 standard for sampling inspection by variables in 1957. It roughly matched the attribute plans in MIL-STD-105A-C, in that the code letters, AQL levels, and OC performance of the plans in MIL-STD-414 were nearly the same as MIL-STD-105A-C. Of course, the variables plans had much lower sample sizes. Figure 3.6 shows the content of MIL-STD-414. The Range method (shown in the grey boxes) simplified hand-calculations by using the range (R) rather than the sample standard deviation (s). However, with modern computers and calculating devices these methods are no longer necessary. The M-method can be used for either single or double specification limits and eliminates the need for the k-Method.

When MIL-STD-105 was updated to version D and E, it destroyed the match between MIL-STD-414 and MIL-STD-105. Commander Gascoigne of the British Navy showed how to restore the match. His ideas were incorporated into the civilian standard ANSI/ASQ Z1.9 in 1980. It now matches OC performance of the plans with the same AQL between the variable plans in ANSI/ASQ Z1.9 and the attribute plans in ANSI/ASQ Z1.4. Therefore it is possible to switch back and forth between an attribute plan in ANSI/ASQ Z1.4 and a variables plan from ANSI/ASQ Z1.9, for the same lot size inspection level and AQL, and keep the same operating characteristic. These plans are recommended for in-house or U.S. domestic trade partners.

These variables sampling schemes are meant to be used for sampling a stream of lots from a supplier. They include normal, tightened, and reduced sampling plans and the same switching rules used by MIL-STD-105E(ANSI/ASQ-Z1.4) Schilling and Neubauer[84]. To use the standard, the supplier and customer companies should agree on an AQL level and adhere to the switching rules. The switching rules must be followed to gain the full benefit of the scheme. Following the rules results in a steep scheme OC curve (approaching the ideal as shown in Figure 2.2 for MIL-STD-105E) with a higher protection level for both the producer and supplier than could be obtained with a single sampling plan with similar sample requirement.

The international derivative of MIL-STD-414 is ISO 3951-1. This set of plans and switching rules is recommended for international trade. The ISO scheme has dropped the plans that use the range (R) as an estimate of process variability (grey boxes in Figure 3.6), and it uses a graphical acceptance criterion for double specification limits in place of the M-method. Using this graphical criterion, a user calculates $\overline{x}$ and the sample standard deviation s from a sample of data, then plots the coordinates $(\overline{x}, s)$ on a curve to see if it falls in the acceptance region.

The function `AAZ19()` in the R package `AQLSchemes` can retrieve the normal, tightened or reduced sampling plans for the variability known or unknown cases from the ANSI/ASQ Z1.9 standard. This function eliminates the need to reference the tables, and provides the required sample size, the acceptability constant (k) and the maximum allowable proportion nonconforming (M). When given the sample data and the specification limits, the function `EPn` in the same package can calculate the estimated proportion non-conforming from sample data as illustrated in section 3.2.

To illustrate the use of these functions, consider the following example of the use of ANSI/ASQ Z1.9. The minimum operating temperature for operation of a device is specified to be 180°F and the maximum operating

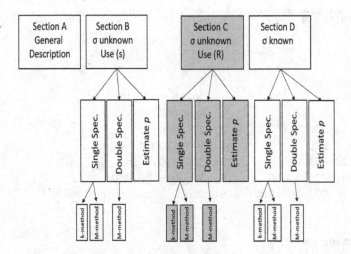

FIGURE 3.6: Content of MIL-STD 414

temperature is 209°F. A lot size $N = 40$ is submitted for inspection with the variability unknown. The AQL is 1%, and the inspection level is II.

The function `AAZ19()` has two required arguments, the first argument `type`, that can take on the values `'Normal'`, `'Tightened'` or `'Reduced'`. It must be supplied to override the default value `'Normal'`. A second argument `stype` can take on the values `'unknown'` or `'known'`. It must be supplied to override the default value `'unknown'`. The function `AAZ19()` is called in the same the way the functions `AASingle()` and `AADouble()` were called. They were illustrated in the last chapter.

The section of R code below illustrates the function call, the interactive queries and answers and the resulting plan. The second argument to the function `AAZ19()` was left out to get the default value.

```
R>library(AQLSchemes)
R>AAZ19('Normal')
MIL-STD-414 ANSI/ASQ Z1.9


What is the Inspection Level?


1: S-3
2: S-4
3: I
4: II
5: III
```

```
Selection: 4

What is the Lot Size?

 1: 2-8                       2: 9-15
 3: 16-25                     4: 26-50
 5: 51-90                     6: 91-150
 7: 151-280                   8: 281-400
 9: 401-500                  10: 501-1200
11: 1201-3200                12: 3201-10,000
13: 10,001-35,000            14: 35,001-150,000
15: 150,001-500,000          16: 500,001 and over


Selection: 4

What is the AQL in percent nonconforming per 100 items?

 1: 0.10    2: 0.15   3: 0.25   4: 0.40   5: 0.65
 6: 1.0     7: 1.5    8: 2.5    9: 4.0   10: 6.5
11: 10

Selection: 6
        n         k         M
5.000000 1.524668 0.033300
```

The result shows that the sampling plan consists of taking a sample of 5 devices from the lot of 40 and comparing the estimated proportion non-conforming to 0.0333.

If the operating temperatures of the 5 sampled devices were: (197,188,184,205, and 201), the EPn() function can be called to calculate the estimated proportion non-conforming as shown in the R code below.

```
R>library(AQLSchemes)
R>sample<-c(197,188,184,205,201)
R>EPn(sample,sided="two",LSL=180,USL=209)
[1] 0.02799209
```

The argument sample in the function call is the vector of sample data values. The argument sided can be equal to "two" or "one", depending on whether there are double specification limits or a single specification limit. Finally, the arguments LSL and USL give the specification limits. If there is only a lower specification limit, change sided="two" to sided="one" and leave out USL. If there is only an upper specification limit leave out LSL in addition to changing the value of sided.

The results of the function call indicate that the estimated proportion non-conforming is 0.02799209, which is less than the maximum tolerable proportion non-conforming = 0.0333; therefore the lot should be accepted. If the sample mean, sample standard deviation, and the sample size have already been calculated and stored in the variables xb, sd, and ns, then the function call can also be given as:

```
R>EPn(sided="two",LSL=180,USL=209,xbar=xb,s=sd,n=ns)
```

The tightened sampling plan for the same inspection level, lot size, and AQL is found with the call:

```
R>library(AQLSchemes)
R>AAZ19('Tightened')
```

Answering the queries the same way as shown above results in the plan:

```
      n        k        M
5.000000 1.645868 0.013400
```

Thus the lot would be rejected under tightened inspection since 0.02799209 > 0.0134.

The reduced sampling plan for the same inspection level, lot size and AQL is found with the call:

```
R>library(AQLSchemes)
R>AAZ19('Reduced')
```

Answering the queries the same way as shown above results in the plan:

```
     n       k       M
4.0000 1.3362 0.0546
```

3.4   Gauge R&R Studies

Repeated measurements of the same part or process output do not always result in exactly the same value. This is called measurement error. It is necessary to estimate the variability in measurement error in order to determine if the measurement process is suitable. Gauge capability or Gauge R&R studies are conducted to estimate the magnitude of measurement error and partition this variability into various sources. The major sources of measurement error are repeatability and reproducibility. Repeatability refers to the error in measurements that occur when the same operator uses the gauge or measuring device to measure the same part or process output repeatedly. Reproducibility refers to the measurement error that occurs due to different measuring conditions such as the operator making the measurement, or the environment where the measurement is made.

A basic gauge R&R study is conducted by having random sample of several operators measure each part or process output in a sample repeatedly. The operators are blinded as to which part they are measuring at the time they measure it.

The results of a gauge R&R study is shown in Table 3.1. In this study, there were three operators. The values represent the measurements of reducing sugar conentration in (g/L) of ten samples of the results of an enzymatic saccharification process for transforming food waste into fuel ethanol. Each operator measured each sample twice.

TABLE 3.1: Results of Gauge R&R Study

Sample	Operator 1	Operator 2	Operator 3
1	103.24	103.16	102.96
2	103.92	103.81	103.76
3	109.13	108.86	108.70
4	108.35	108.11	107.94
5	105.51	105.06	104.84
6	106.63	106.61	106.60
7	109.29	108.96	108.84
8	108.76	108.39	108.23
9	108.03	107.86	107.72
10	106.61	106.32	106.21
1	103.56	103.26	103.01
2	103.86	103.80	103.75
3	109.23	108.79	108.75
4	108.29	108.24	107.99
5	105.53	105.11	104.80
6	106.65	106.57	106.55
7	109.28	109.12	109.03
8	108.72	108.43	108.27
9	108.11	107.84	107.79
10	106.77	106.23	106.13

An analysis of variance is used to to estimate σ_p^2, the variance among different parts (or in this study samples); σ_o^2, the variance among operators; σ_{po}^2, the variance among part by operator; and σ_r^2, the variance due to repeat measurements on one part (or sample in this study) by one operator. The gauge repeatability variance is defined to be σ_r^2, and the gauge reproducibility variance is defined to be $\sigma_o^2 + \sigma_{po}^2$.

In the Rcode and output below, the measurement data from Table 3.1 are entered row by row into a single vector d, where operator changes fastest, sample second fastest, and the repeat measurements third. Next, the gageRRDesign() function in the `qualityTools` package is used to create the indicator variables for Operators, Parts, and Measurements in the design matrix gRR. The statement `response(gRR)<-d` adds the measurment data in `d` to the design , and the `gageRR()` function/indexgageRR() function, qualityTools package produces the Analysis of Variance.

Below the Analysis of Variance are the estimates of: $\sigma_r^2 = 0.00478$ (repeatability), $\sigma_o^2 = 0.03621$ (Operator), $\sigma_{po}^2 = 0.00733$ (Operator:Part), and $\sigma_p^2 = 4.47552$ (Part to Part). The reproducibility is $\sigma_o^2 + \sigma_{po}^2 = 0.04354$. The measurment error variance is then $\sigma_{gauge}^2 = 0.04832$ (totalRR), which is the sum of the repeatability and reproducibility.

```
R>library(qualityTools)
R>design=gageRRDesign(Operators=3, Parts=10 ,
  Measurements=2, randomize=FALSE)
R>#set the response
R>response(design)=c(103.24,103.16,102.96,103.92,103.81,
                     103.76,109.13,108.86,108.70,108.35,
                     108.11,107.94,105.51,105.06,104.84,
                     106.63,106.61,106.60,109.29,108.96,
                     108.84,108.76,108.39,108.23,108.03,
                     107.86,107.72,106.61,106.32,106.21,
                     103.56,103.26,103.01,103.86,103.80,
                     103.75,109.23,108.79,108.75,108.29,
                     108.24,107.99,105.53,105.11,104.80,
                     106.65,106.57,106.55,109.28,109.12,
                     109.03,108.72,108.43,108.27,108.11,
                     107.84,107.79,106.77,106.23,106.13)
R>gA<-gageRR(design)
R>plot(gA)


AnOVa Table -  crossed Design
              Df Sum Sq Mean Sq  F value                 Pr(>F)
Operator       2   1.49   0.744  155.740  < 0.0000000000000002
Part           9 241.85  26.873 5627.763  < 0.0000000000000002
Operator:Part 18   0.35   0.019    4.072              0.000346
Residuals     30   0.14   0.005

Operator      ***
Part          ***
Operator:Part ***
Residuals
---
Signif. codes:  0 '***' 0.001 '**' 0.01 '*' 0.05 '.' 0.1 ' ' 1


----------
```

```
Gage R&R
                 VarComp VarCompContrib  Stdev StudyVar
totalRR          0.04832        0.01068 0.2198    1.319
 repeatability   0.00478        0.00106 0.0691    0.415
 reproducibility 0.04354        0.00963 0.2087    1.252
   Operator      0.03621        0.00800 0.1903    1.142
   Operator:Part 0.00733        0.00162 0.0856    0.514
Part to Part     4.47552        0.98932 2.1155   12.693
totalVar         4.52384        1.00000 2.1269   12.762
                 StudyVarContrib
totalRR                  0.1033
 repeatability           0.0325
 reproducibility         0.0981
   Operator              0.0895
   Operator:Part         0.0403
Part to Part             0.9946
totalVar                 1.0000


---
 * Contrib equals Contribution in %
 **Number of Distinct Categories (truncated
 signal-to-noise-ratio) = 13
```

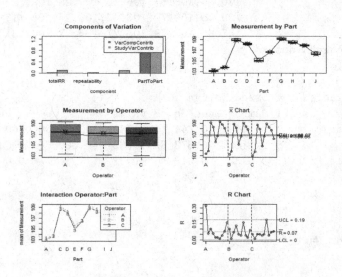

FIGURE 3.7: Gauge R&R Plots

The graph produced by the `plot(gA)` statement can identify any outliers that may skew the results. From the boxplots in the upper right and middle left, it can be seen that variability of measurements on each part are reasonably consistent and operators are consistent with no apparent outliers.

From the plots on the top left and middle right, it can be seen that σ^2_{gauge} is small relative to the σ^2_p. Generally, the gauge or measuring instrument is considered to be suitable if the process to tolerance $P/T = \frac{6 \times \sigma_{gauge}}{USL-LSL} \leq$ 0.10 where $\sigma_{gauge} = \sqrt{\sigma^2_{gauge}}$ and USL, and LSL are the upper and lower specification limits for the part being measured.

If the P/T ratio is greater than 0.1, looking at the $\sigma^2_{repeatability}$ and $\sigma^2_{reproducibility}$ gives some information about how to improve the measuring process. If $\sigma^2_{repeatability}$ is the largest portion of measurement error it would indicate that the gauge or measuring device is inadequate since this variance represents the variance of repeat measurements of the same part by the same operator with the same gauge. If $\sigma^2_{reproducibility}$ is the largest portion of measurement error, and if the plots on the middle and bottom left showed large variability in operator averages or inconsistent trends of measurements across parts for each operator, then perhaps better training of operators could reduce σ^2_o and σ^2_{po} thereby reducing σ^2_{gauge}.

3.5   Additional Reading

Chapters 2 and 3 have presented an introduction to lot-by-lot attribute and variable sampling plans and AQL based attribute and variable sampling schemes. Additional sampling plans exist for rectification sampling, accept-on-zero sampling plans, and continuous sampling plans that are useful when there is a continuous flow of product that can't naturally be grouped. Books by Shmueli[87] and Schilling and Neubauer[84] present more information on these topics.

Section 3.4 presented an introduction to measurement analysis. Again there is a more comprehensive coverage of this topic in Burdick, Borror, and Montgomery[13].

3.6   Summary

This chapter has discussed variables sampling plans and schemes. The major advantage to variables sampling plans over attribute plans is the same protection levels with reduced sample sizes. Table 3.1 (patterned after one presented by Schilling and Neubauer[84] shows the average sample numbers for various plans that are matched to a single sampling plan for attributes with $n = 50$, $c = 2$. In addition to reduced sample sizes, variable plans provide information like the mean and estimated proportion defective below the lower specification limit and above the upper specification limit. This information can be valuable to the producer in correcting the cause of rejected lots and improving the process to produce at the AQL level or better.

TABLE 3.2: Average Sample Numbers for Various Plans

Plan	Average Sample Number
Single Attributes	50
Double Attributes	43
Multiple Attributes	35
Variables (σ unknown)	27
Variables (σ known	12

When a continuous stream of lots is being sampled, the published schemes with switching rules are more appropriate. They provide better protection for producer and consumer at a reduced average sample number. The variables plans and published tables described in this chapter are based on the assumption that the measured characteristic is normally distributed.

That being said, the need for any kind of acceptance sampling is dependent on the consistency of the supplier's process. If the supplier's process is consistent (or in a state of statistical control) and is producing defects or nonconformities at a level that is acceptable to the customer, Deming[21] pointed out that no inspection is necessary or cost effective. On the other hand, if the supplier's process is consistent but producing defects or nonconformities at a level that is too high for the customer to tolerate, 100% inspection should always be required. This is because the number (or proportion) nonconforming in a random sample from the lot is uncorrelated with the proportion nonconforming in the remainder of the lot. This can be demonstrated with the following simple example.

If the producer's process is stable and producing 3% nonconforming and delivering lots of 200 items to the customer, then the sampling results using an attribute single sampling plan a sample size of $n = 46$ and an acceptance number of $c = 3$ can be simulated with the following R code. **ps** represents the proportion defective in the sample, and **pr** represents the proportion defective in the remainder of the lot.

```
R># Lot size N=200 with an average 3% defective
R>p<-rbinom(50,200,.03)
R>r<-seq(1:50)
```

```
R># This loop simulates the number non-conforming in a sample
R>#  of 46 items from each of the simulated lots
R>for (i in 1:length(p)) {
R>  r[i]<-rhyper(1,p[i],200-p[i],46)
R> }
R># this statement calculates the proportion non-conforming
R># in each lot
R>ps<-r/46
R>#This statement calculates the proportion non-conforming
R>pr<-(p-r)/154
R># in the unsampled portion of each lot
R>plot(pr~jitter(ps,1),xlab='Proportion nonconforming in
   Sample of 46', ylab='Proportion nonconforming in
   remainder of Lot')
R>abline(h=(.03*46)/46,lty=2)
R>cor(ps,pr)
```

Figure 3.8 is a plot of the simulated proportion nonconforming in the sample of 46 versus the proportion nonconforming in the remainder of the lot of $200 - 46$. It can be seen that there is no correlation. When the process is stable, rejecting lots with more than 3 defectives in the sample and returning them to the producer will not change the overall proportion of defects the customer is keeping.

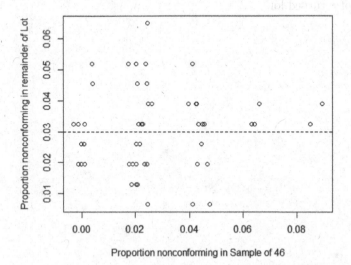

FIGURE 3.8: Simulated between the proportion nonconforming in a sample and the proportion nonconforming in the remainder of the lot

The same relationship will be true for attribute or variable sampling. This is exactly the reason that in 1980 Ford Motor Company demanded that their suppliers demonstrate their processes were in a state of statistical control with a proportion of nonconforming items at a level they could tolerate. Under those circumstances, no incoming inspection was necessary. Ford purchased a large enough share of their supplier's output to make this demand. Following the same philosophy, the U.S. Department of Defense issued MIL-STD-1916 in 1996. This document stated that sampling inspection itself was an inefficient way of demonstrating conformance to the requirements of a contract, and that defense contractors should instead use process controls and statistical control methods. Again the large volume of supplies procured by the Department of Defense allowed them to make this demand of their suppliers. For smaller companies, who do not have that much influence on their suppliers, acceptance sampling of incoming lots may be the only way to assure themselves of adequate quality of their incoming components.

Chapter 4 will consider statistical process control and process improvement techniques that can help a company achieve stable process performance at an acceptable level of nonconformance. These methods should always be used when the producer is in-house. In cases where the producer is external to the customer, the use of sampling schemes like ANSI/ASQ-Z1.9 can encourage producers to work on making their processes more consistent with a low proportion of non-conformance. This is the case because a high or inconsistent level of nonconformance from lot to lot will result in a switch to tightened inspection using a sampling scheme. This increases the producer's risk and the number of returned lots.

3.7 Exercises

1. Run all the code examples in the chapter, as you read.

2. Show that $(\bar{x} - LSL)/\sigma > k$ (the inequality in Equation 3.2), implies that $(\bar{x} - \mu_{RQL})/(\sigma/\sqrt{n}) > k\sqrt{n} + (LSL - \mu_{RQL})/(\sigma/\sqrt{n})$.

3. What is the distribution of $(\bar{x} - \mu_{RQL})/(\sigma/\sqrt{n})$, when x follows a $N(\mu_{RQL}, \sigma)$?

4. The density of a plastic part used in a mobile phone is required to be at least 0.65g/cm^3. The parts are supplied in lots of 5000. The AQL and LTPD are 0.01 and 0.05, and $\alpha{=}.05$, $\beta{=}.10$.

   (a) Find an appropriate variables sampling plan (n and k) assuming the density is normally distributed with a known standard deviation σ.

   (b) Find an appropriate variables sampling plan assuming the standard deviation is unknown.

   (c) Find an appropriate attributes sampling plan (n and c).

   (d) Are the OC curves for the three plans you have found similar?

   (e) What are the advantages and disadvantages of the variables sampling plan in this case.

5. A supplier of components to a small manufacturing company claims that they can send lots with no more than 1% defective components, but they have no way to prove that with past data. If they will agree to allow the manufacturing company to conduct inspection of incoming lots using an ANSI/ASQ-Z1.9 sampling scheme (and return rejected lots), how might this motivate them to be sure they can meet the 1% AQL?

6. The molecular weight of a polymer product should fall within $LSL{=}2100$ and $USL{=}2350$, the AQL=1%, and the RQL=8% with $\alpha = 0.05$, and $\beta = 0.10$. It is assumed to be normally distributed.

   (a) Assuming the standard deviation is known to be $\sigma = 60$, find and appropriate variables sampling plan.

   (b) Find the appropriate variables sampling plan for this situation if the standard deviation σ is unknown.

   (c) If a sample of the size you indicate in (b). was taken from a lot and $\bar{x}$ was found to be 2221, and the sample standard deviation s was found to be 57.2, would you accept or reject the lot?

7. If the lot size is N=60, and the AQL=1.0%, use the `AAZ19()` function to find the ANSI/ASQ Z1.9 sampling plan under Normal, Tightened, and Reduced inspection for the variability unknown case.

(a) Find the OC curves for the Normal and Tightened plans using the
OCvar() funtion in the AcceptanceSampling package, and retrieve
the OC values as shown in Section 3.1.2.2.

(b) Use the OC values for the Normal and Tightened plans and equations
2.12 and 2.13 in Chapter 2 to find the OC curve for the scheme that
results from following the ANSI/ASQ Z1.9 switching rules.

(c) Plot all three OC curves on the same graph.

(d) If the specification limits were LSL=90, USL=130, and a sample of 7
resulted in the measurements: 123.13, 103.89, 117.93, 125.52, 107.79,
113.06, 100.19, would you accept or reject the lot under Normal inspection?

4

Shewhart Control Charts in Phase I

4.1 Introduction

Statistical Quality Control (SQC) consists of methods to improve the quality of process outputs. Statistical Process Control (SPC) is a subset of Statistical Quality Control (SQC). The objective for SPC is to specifically understand, monitor, and improve processes to better meet customer needs.

In Chapter 2 and 3, the MIL-STD documents for acceptance sampling were described. In 1996, the U.S. Department of Defense realized that there was an evolving industrial product quality philosophy that could provide defense contractors with better opportunities and incentives for improving product quality and establishing a cooperative relationship with the Government. To this end MIL-STD-1916 was published. It states that process controls and statistical control methods are the preferable means of preventing nonconformances; controlling quality; and generating information for improvement. Additionally, it emphasizes that sampling inspection by itself is an inefficient industrial practice for demonstrating conformance to the requirements of a contract.

Later civilian standards have provided more detailed guidelines, and should be followed by all suppliers whether they supply the Government or private companies. ASQ/ANSI/ISO 7870-2:2013 establishes a guide for the use and understanding of the Shewhart control chart approach for statistical control of a process. ISO 22514-1:2014—Part 1: provides general principles and concepts regarding statistical methods in process management using capability and performance studies. ASQ/ANSI/ISO 7870-4:2011 provides statistical procedures for setting up cumulative sum (cusum) schemes for process and quality control using variables (measured) and attribute data. It describes general-purpose methods of decision-making using cumulative sum (cusum) techniques for monitoring and control. ASQ/ANSI/ISO 7870-6:2016 Control charts—Part 6: EWMA control charts: Describes the use of EWMA Control Charts in Process Monitoring. These documents can be accessed at https://asq.org/quality-press/, and all the methods will be described in detail in this chapter and chapter 6.

The output of all processes, whether they are manufacturing processes or processes that provide a service of some kind, are subject to variability. Variability makes it more difficult for processes to generate outputs that fall within desirable specification limits. Shewhart[86] designated the two causes for variability in process outputs to be common causes and special or assignable causes. Common causes of variability are due to the inherent nature of the process. They can't be eliminated or reduced without changing the actual process. Assignable causes for variability, on the other hand, are unusual disruptions to normal operation. They should be identified and removed in order to reduce variability and make the process more capable of meeting the specifications.

Control charts are statistical tools. Their use is the most effective way to distinguish between common and assignable cause for variability when monitoring process output in real time. Although control charts alone cannot reduce process variability, they can help to prevent over-reaction to common causes for variability (which may make things worse), and help to prevent ignoring assignable cause signals. When the presence of an assignable cause for variability is recognized, knowledge of the process can lead to adjustments to remove this cause and reduce the variability in process output.

To illustrate how control charts distinguish between common and assignable causes, consider the simple example described by Joiner[46]. Eleven year old Patrick Nolan needed a science project. After talking with his father, a statistician, he decided to collect some data on something he cared about; his school bus. He recorded the time the school bus arrived to pick him up each morning, and made notes about anything he considered to be unusual that morning. After about 5 weeks, he made the chart shown in Figure 4.1 that summarized the information he collected.

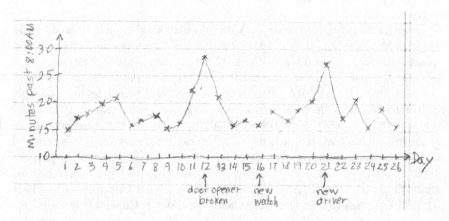

FIGURE 4.1: Patrick's Chart (Source B.L. Joiner *Fourth Generation Management*, McGraw Hill, NY, ISBN0-07032715-7)

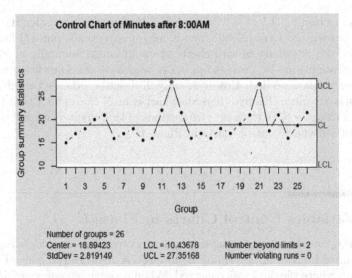

FIGURE 4.2: Control Chart of Patrick's Data

A control chart of the number of minutes past 8:00AM that the bus arrived is shown in Figure 4.2. In this figure, the upper control limit (UCL) is three standard deviations above the mean, and the lower control limit (LCL) is three standard deviations below the mean. The two points in red are above the upper control limit, and they correspond to the day when Patrick noted that the school bus door opener was broken, and the day when there was a new driver on the bus. The control chart shows that these two delayed pickup times were due to special or assignable causes (which were noted as unusual by Patrick). The other points on the chart are within the control limits and appear to be due to common causes. Common causes like normal variation in amount of traffic, normal variation in the time the bus driver started on the route, normal variation in passenger boarding time at previous stops, and slight variations in weather conditions are always present and prevent the pickup times from being exactly the same each day.

There are different types of control charts, and two different situations where they are used (Phase I, and Phase II (see Chakraborti[18])). Most textbooks describe the use of Shewhart control charts in what would be described as Phase II process monitoring. They also describe how the control limits are calculated by hand using tables of constants. The constants used for calculating the limits can be found in the vignette SixSigma::ShewhartConstants in the R package SixSigma. There are also functions ss.cc.getd2(), ss.cc.getd3(), and ss.cc.getc4() in the SixSigma package for retrieving these and other control chart constants. Examples of their use are shown by Cano et. al.[16]. In addition, tables of the constants for computing control chart limits, and procedure for using them to calculate the limits by hand are shown in Section 6.2.3.1 of the online NIST Engineering Statistics Handbook (https://www.itl.nist.gov/div898/handbook/pmc/section3/pmc321.htm)[1].

Sometimes Shewhart control charts are maintained manually in Phase II real time monitoring by process operators who have the knowledge to make adjustments when needed. However, we will describe other types of control charts that are more effective than Shewhart control charts for Phase II monitoring in Chapter 6. In the next two sections of this chapter, we will describe the use of Shewhart control charts in Phase I.

4.2   Variables Control Charts in Phase I

In Phase I, control charts are used on retrospective data to calculate preliminary control limits and determine if the process had been in control during the period where the data was collected. When assignable causes are detected on the control chart using the historical data, an investigation is conducted to find the cause. If the cause is found, and it can be prevented in the future, then the data corresponding to the out of control points on the chart are eliminated and the control limits are recalculated.

This is usually an iterative process and it is repeated until control chart limits are refined, a chart is produced that does not appear to contain any assignable causes, and the process appears to be operating at an acceptable level. The information gained from the Phase I control chart is then used as the basis of Phase II monitoring. Since control chart limits are calculated repeatedly in Phase I, the calculations are usually performed using a computer. This chapter will illustrate the use of R to calculate control chart limits and display the charts.

Another important purpose for using control charts in Phase I is to document the process knowledge gained through an Out of Control Action Plan (OCAP). This OCAP is a list of the causes for out of control points. This list is used when monitoring the process in Phase II using the revised control limits determined in Phase I. When out of control signals appear on the chart in Phase II, the OCAP should give an indication of what can be adjusted to bring the process back into control.

The Automotive Industry Action Group[96] recommends preparatory steps be taken before variable control charts can be used effectively. These steps are summarized in the list below.

1. **Establish an environment suitable for action.** Management must provide resources and support actions to improve processes based on knowledge gained from use of control charts.

2. **Define the process.** The process including people, equipment, material, methods, and environment must be understood in relation to the upstream

suppliers and downstream customers. Methods such as flowcharts, SIPOC diagrams (to be described in the next chapter), and cause-and-effect diagrams described by later in this chapter are useful here.

3. **Determine characteristics to be charted.** This step should consider customer needs, current and potential problems, and correlations between characteristics. For example, if a characteristic of the customer's need is difficult to measure, track a correlated characteristic that is easier to measure.

4. **Define the measurement system.** Make sure the characteristic measured is defined so that it will have the same meaning in the future and the precision and accuracy of the measurement is predictable. Tools such as Gauge R&R Studies described in Chapter 3 are useful here.

Data for Shewhart control charts are gathered in subgroups. Subgroups, properly called *rational subgroups*, should be chosen so that the chance for variation among the process outputs within a subgroup is small and represents the inherent process variation. This is often accomplished by grouping process outputs generated consecutively together in a subgroup, then spacing subgroups far enough apart in time to allow for possible disruptions to occur between subgroups. In that way, any unusual or assignable variation that occurs between groups should be recognized with the help of a control chart. If a control chart is able to detect out of control signals, and the cause for these signals can be identified, it is an indication that the rational subgroups were effective.

The data gathering plan for Phase I studies are defined as follows:

1. **Define the subgroup size** Initially this is a constant number of 4 or 5 items per each subgroup taken over a short enough interval of time so that variation among them is due only to common causes.

2. **Define the Subgroup Frequency** The subgroups collected should be spaced out in time, but collected often enough so that they can represent opportunities for the process to change.

3. **Define the number of subgroups** Generally 25 or more subgroups are necessary to establish the characteristics of a stable process. If some subgroups are eliminated before calculating the revised control limits due to discovery of assignable causes, additional subgroups may need to be collected so that there are at least 25 subgroups used in calculating the revised limits.

4.2.1 Use of $\overline{X}$-R charts in Phase I

The control limits for the $\overline{X}$-chart are $\overline{\overline{X}} \pm A_2\overline{R}$, and the control limits for the R-chart are $D_3\overline{R}$ and $D_4\overline{R}$. The constants A_2, D_3, and D_4 are indexed by the rational subgroup size and are again given in the vignette

SixSigma::ShewhartConstants in the R package SixSigma. They can also be found in Section 6.2.3.1 of the online NIST Engineering Statistics Handbook. These constants are used when calculating control limits manually. They are not needed when the control charts are created using R functions. We will illustrate the using the qcc() function in the R package qcc for creating $\overline{X} - R$ Charts in Phase I using data from Mitra[70]. He presents the following example in which subgroups of 5 were taken from a process that manufactured coils. The resistance values in ohms was measured on each coil, and 25 subgroups of data were available. The data is shown in Table 4.1.

4.2.2 $\overline{X}$ and R Charts

To begin, this historical data is entered into a computer file so that it can be analyzed. For example, Figure 4.3 shows the first three lines of a .csv (comma separated) file containing this data. The first line contains names for the variables in each subgroup.

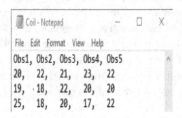

FIGURE 4.3: Coil.csv file

The data is read into an R data frame using the commands shown below, and the qcc() function in the R package qcc is used to create the R-chart shown in Figure 4.4.

```
R>CoilG <- read.table("Coil.csv", header=TRUE,sep=",",
   na.strings="NA",dec=".",strip.white=TRUE)
R>library(qcc)
R>qcc(CoilG, type="R")
```

In Figure 4.4, the range for subgroup 3 is out of the limits for the R-chart, indicating an assignable cause was present. Including this out-of-control subgroup in the calculations increases the average range to $\overline{R} = 3.48$, and will broaden the control limits for both the $\overline{X}$ chart and R-chart and reduce the chances of detecting other assignable causes.

When the cause of the wide range in subgroup 3 was investigated, past information showed that was the day that a new vender of raw materials and components was used. The quality was low, resulting in the wide variability in coil resistance values produced on that day. Based on that fact, management instituted a policy to require new vendors to provide documentation that their

TABLE 4.1: Coil Resistance

Subgroup	Measured Ohms				
1	20	22	21	23	22
2	19	18	22	20	20
3	25	18	20	17	22
4	20	21	22	21	21
5	19	24	23	22	20
6	22	20	18	18	19
7	18	20	19	18	20
8	20	18	23	20	21
9	21	20	24	23	22
10	21	19	20	20	20
11	20	20	23	22	20
12	22	21	20	22	23
13	19	22	19	18	19
14	20	21	22	21	22
15	20	24	24	23	23
16	21	20	24	20	21
17	20	18	18	20	20
18	20	24	22	23	23
19	20	19	23	20	19
20	22	21	21	24	22
21	23	22	22	20	22
22	21	18	18	17	19
23	21	24	24	23	23
24	20	22	21	21	20
25	19	20	21	21	22

process could meet the required standards. With this in mind, subgroup 3 is not representative of the process after implemention of this policy, so it should be removed before calculating and displaying the control charts.

In the R code shown below, subgroup 3 is removed from the updated data frame Coilm3 and the $\overline{X}$ chart shown in Figure 4.5 was produced.

```
R># Eliminate subgroup 3
R>library(qcc)
R>Coilm3<-CoilG[-c(3), ]
R>qcc(Coilm3, type="xbar")
```

When investigating the causes for the the high averages for subgroups 15 and 23, and the low average for subgroup 22, it was found that the oven temperature was too high for subgroup 22, and the wrong die was used for subgroup 23, but no apparent reason for the high average for subgroup 15 could be found.

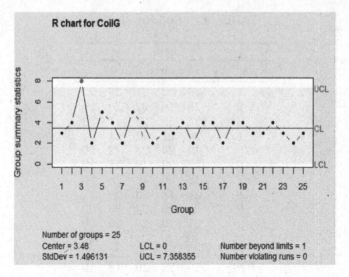

FIGURE 4.4: *R*-chart for Coil Resistance

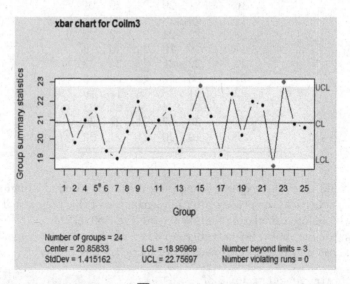

FIGURE 4.5: $\overline{X}$-chart for Coil Resistance

Assuming high oven temperatures and use of the wrong die can be avoided in the future, the control chart limits were recomputed eliminating subgroups 22 and 23. This resulted in the upper and lower control limits for the *R*-chart (LCL=0, UCL=6.919) and the $\overline{X}$ chart (LCL=18.975, UCL=22.753). However, one subgroup (15) remains out of the control limits on the $\overline{X}$ chart, and there are less than 25 total subgroups of data used in calculating the limits.

The next step would be to eliminate the data from subgroup 15, collect additional data, and recompute the limits and plot the chart again. It may take many iterations before charts can be produced that do not exhibit any assignable causes. As this is done, the assignable causes discovered in the process should be documented into an *Out of Control Action Plan* or OCAP. For example, in the two iterations described above, three assignable causes were discovered. An OCAP based on just these three causes mentioned may look something like the list shown below.

OCAP
Out of Control on R Chart–High within subgroup variability

1. Verify that the vendor of raw material and component has documented their ability supply within specs. If not, switch vendors.

Out of control on $\overline{X}$ chart

1. Check to make sure the proper die was used. If not, switch to the proper die.

2. Check to make sure oven temperature was set correctly. If not, set correctly for the next production run.

In practice, considerable historical data from process operation may already be routinely collected and stored in data bases. The R package `readr` has several functions that can be used to load plain text rectangular files into R data frames (see Wickhamand and Grolemund[102]). Most large companies maintain a Quality Information System to be described later in this chapter. As long as the data that is retrieved can be formatted as rational subgroups (with only common cause variability within subgroups), this is a good source of historical data for Phase I control chart applications.

The discovery of assignable causes for out of control signals in real applications can sometimes be a difficult task. If there is not an obvious cause, then the Cause-and-Effect (or Ishakawa) Diagram, presented later in this in Chapter, is useful for organizing and recording ideas for possible assignable causes.

Frequently an out of control signal is caused by a combination of changes to several factors simultaneously. If that is the case, testing changes to one-factor-at a time is inefficient and will never discover the cause. Several factors can be tested simultaneously using Experimental Design plans. Some of these methods will be presented in the next chapter and more details can be found in Lawson[57].

Since the control charts produced with the data in Table 4.1 detected out of control signals, it is an indication that the rational subgroups were effective.

There was less variability within the subgroups than from group to group, and therefore subgroups with high or low averages were identified on the $\overline{X}$ chart as potential assignable causes. If data is retrieved from a historical database, and no assignable causes are identified on either the $\overline{X}$ or R charts and subgroup means are close to the center line of the $\overline{X}$ chart, it would indicate that the subgroups have included more variability within the subgroups than from subgroup to subgroup. These subgroups will be ineffective in identifying assignable causes.

Notice that the control limits for the $\overline{X}$ and R charts described in Section 6.2.3.1 of the online NIST Engineering Statistics Handbook (https://www.itl.nist.gov/div898/handbook/pmc/section3/pmc321.htm)[1], and computed by the `qcc` function as shown in Figure 4.3, have no relationship with the specification limits described in Chapter 3. Specification limits are determined by a design engineer or another familiar with the customer needs (that may be the next process). The control limits for the control chart are calculated from observed data, and they show the limits of what the process is currently capable of. The `Center` (or $\overline{\overline{X}}$) and the `StDev` (or standard deviation) shown in the group of statistics at the bottom of the chart in Figure 4.5 are estimates of the current process average and standard deviation. Specification limits should never be added to the control charts because they will be misleading.

The goal in Phase I is to discover and find ways to eliminate as many assignable causes as possible. The revised control chart limits and OCAP will then be used in Phase II to keep the process operating at the optimal level with minimum variation in output.

4.2.3   Interpreting Charts for Assignable Cause Signals

In addition to just checking individual points against the control limits to detect assignable causes, an additional list of indicators that should be checked is shown below.

1. A point above or below the upper control limit or below the lower control limit

2. Seven consecutive points above (or below) the center line

3. Seven consecutive points trending up (or down)

4. Middle one-third of the chart includes more than 90% or fewer than 40% of the points after at least 25 points are plotted on the chart

5. Obviously non-random patterns

These items are based on the Western Electric Rules proposed in the 1930s. If potential assignable cause signals are investigated whenever any one of the indicators are violated (with the possible exception of number 3.), it will make the control charts more sensitive for detecting assignable causes. It has been shown that the trends of consecutive points, as described in 3, can sometimes occur by random chance, and strict application of indicator 3 may lead to false positive signals and wasted time searching for assignable causes that are not present.

4.2.4 $\overline{X}$ and s Charts

The $\overline{X}$ and s Charts described in the online NIST Engineering Statistics Handbook can also be computed by the qcc() function by changing the option type="R" to type="S" in the function call for the R-chart. When the subgroup size is 4 or 5, as normally recommended in Phase I studies, there is little difference between $\overline{X}$ and s charts and $\overline{X}$ and R Charts, and it doesn't matter which is used.

4.2.5 Variable Control Charts for Individual Values

In some situations there may not be a way of grouping observations into rational subgroups. Examples are continuous chemical manufacturing processes, or administrative processes where the time to complete repetitive tasks (i.e., cycle time) are recorded and monitored. In this case a control chart of individual values can be created. Using qcc function and the option type="xbar.one" a control chart of individual values is made. The block of R code below shows how the chart shown in Figure 4.2 was made.

```
R>minutes<-c(15,17,18,20,21,16,17,18,15.5,16,22,28,21.5,16,
    17,16,18,17,19,21,27.5,17.5,21,16,18.75,21.5)
R>library(qcc)
R>qcc(minutes, type="xbar.one", std.dev="MR",
    title="Control Chart of Minutes after 8:00AM")
```

The option std.dev="MR" instructs the function to estimate the standard deviation by taking the average moving range of two consecutive points (i.e., $(|17-15|+|18-17|,\dots|21.5-18.75|)/25$). The average moving range is then scaled by dividing by d_2 (from the factors for control charts with subgroup size $n = 2$). This is how the Std.Dev=2.819149 shown at the bottom of Figure 4.2 was obtained. This standard deviation was used in constructing the control limits $UCL = 18.89423 + 3 \times 2.819149 = 27.35168$ and $LCL = 18.89423 - 3 \times 2.819149 = 10.43678$. std.dev="MR" is the default, and leaving it out will not change the result.

The default can be changed by specifying std.dev="SD". This specification will use the sample standard deviation of all the data as an estimate of the standard deviation. However, this is usually not recommended, because any assignable causes in the data will inflate the sample standard deviation, widen the control limits, and possibly conceal the assignable causes within the limits.

4.3   Attribute Control Charts in Phase I

While variable control charts track measured quantities related to the quality of process outputs, attribute charts track counts of nonconforming items. Attribute charts are not as informative as variables charts for Phase I studies. A shift above the upper or below the lower control limit or a run of points above or below the center line on a variables chart may give a hint about the cause. However, a change in the number of nonconforming items may give no such hint. Still attribute charts have value in Phase I.

In service industries and other non-manufacturing areas, counted data may be abundant but numerical measurements rare. Additionally, many characteristics of process outputs can be considered simultaneously using attribute charts.

Section 6.3.3 of the online NIST Engineering Statistics Handbook describes p, np, c, and u charts for attribute data. All of these attribute charts can be made with the `qcc()` function in the R package `qcc`. The formulas for the control limits for the p-chart are:

$$UCL = \overline{p} + 3\sqrt{\frac{\overline{p}(1 - \overline{p})}{n}}$$

$$LCL = \overline{p} - 3\sqrt{\frac{\overline{p}(1 - \overline{p})}{n}}, \tag{4.1}$$

where n is the number of items in each subgroup. The control limits will be different for each subgroup if the subgroup sizes n are not constant. For example, the R code below produces a p chart with varying control limits using the data in Table 14.1 on page 189 of [19]. When the lower control limit is negative, it is always set to zero.

```
R>library(qcc)
R>d<-c(3,6,2,3,5,4,1,0,1,0,2,5,3,6,2,4,1,1,6,5,6,4,3,4,
   1,2,5,0,0)
R>n<-c(48,45,47,51,48,47,48,50,46,45,47,48,50,50,49,46,
   50,52,48,47,49,49,51,50,48,47,47,49,49)
R>qcc(d,sizes=n,type="p")
```

In the code, the vector d is the number of nonconforming, and the vector n is the sample size for each subgroup. The first argument to qcc is the number of nonconforming, and the second argument sizes= designates the sample sizes. It is a required argument for type="p", type="np", or type="u" attribute charts. The argument sizes can be a vector, as shown in this example, or as a constant.

4.3.1 Use of a p Chart in Phase I

The qcc package includes data from a Phase I study using a p chart taken from Montgomery[72]. The study was on a process to produce cardboard cans for frozen orange juice. The plant management requested that the use of control charts be implemented in effort to improve the process. In the Phase I study, 30 subgroups of 50 cans each were initially inspected at half-hour intervals and classified as either conforming or nonconforming. The R code below taken from the qcc package documentation makes the data available. The first three lines of the data frame orangejuice are shown below the code.

```
R>library(qcc)
R>data(orangejuice)
R>attach(orangejuice)
R>orangejuice
    sample  D size trial
1        1 12   50  TRUE
2        2 15   50  TRUE
3        3  8   50  TRUE
         .
         .
         .
```

The column sample in the data frame is the subgroup number, D is the number of nonconforming cans in the subgroup, size is the number of cans inspected in the subgroup, and trial is an indicator variable. Its value is TRUE for each of the 30 subgroups in the initial sample (there are additional subgroups in the data frame). The code below produces the initial p chart shown in Figure 4.6.

```
R>with(orangejuice,
    qcc(D[trial], sizes=size[trial], type="p"))
```

In this code the [trial] (after D and size), restricts the data to the initial 30 samples where trial=TRUE.

In this Figure, the proportion nonconforming for subgroups 15 and 23 fell above the upper control limit. With only, conforming-nonconforming information available on each sample of 50 cans, it could be difficult to determine the cause of a high proportion nonconforming. A good way to gain insight about the possible cause of an out of control point is to classify the nonconforming items. For example, if the 50 nonconforming items in sample 23 were classified into 6 types of nonconformites. A Pareto diagram, described later in this chapter, could then be constructed.

After a Pareto diagram was constructed, and it showed that the sealing failure and adhesive failure represented more than 65% of all defects. These defects usually reflect operator error. This sparked an investigation of what operator was on duty at that time. It was found that the half-hour period

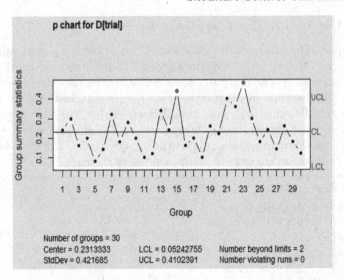

FIGURE 4.6: p chart of the number nonconforming

when subgroup 23 was obtained occurred when a temporary and inexperienced operator was assigned to the machine; and that could account for the high fraction nonconforming.

A similar classification of the nonconformities in subgroup 15 led to the discovery that a new batch of cardboard stock was put into production during that half hour period. It was known that introduction of new batches of raw material sometimes caused irregular production performance. Preventative measures were put in place to prevent use of inexperienced operators and untested raw materials in the future.

Subsequently, subgroups 15 and 23 were removed from the data, and the limits were recalculated as $\bar{p} = .2150$, $UCL = 0.3893$, and $LCL = 0.0407$. Subgroup 21 ($p = .40$) now falls above the revised UCL.

There were no other obvious patterns, and at least with respect to operator controllable problems, it was felt that the process was now stable.

Nevertheless, the average proportion nonconforming cans is 0.215, which is not an acceptable quality level (AQL). The plant management agreed, and asked the engineering staff to analyze the entire process to see if any improvements could be made. Using methods like those to be described in Chapter 5, the engineering staff found that several adjustments could be made to the machine that should improve its performance.

After making these adjustments, 24 more subgroups (numbers 31-54) of $n=50$ were collected. The R code below plots the fraction nonconforming for these additional subgroups on the chart with the revised limits. The result is shown in Figure 4.7.

```
R>library(qcc)
R># remove out-of-control points (see help(orangejuice)
   for the reasons)
R>inc <- setdiff(which(trial), c(15,23))
R>qcc(D[inc], sizes=size[inc], type="p", newdata=D[!trial],
   newsizes=size[!trial])
R>detach(orangejuice)
```

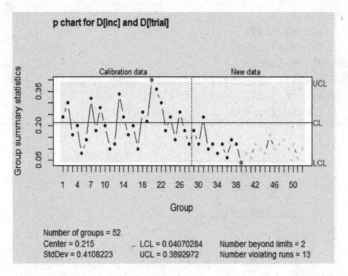

FIGURE 4.7: Revised limits with additional subgroups

The statement `inc <- setdiff(which(trial), c(15,23))` creates a vector (`inc`) of the subgroup numbers, where `trial=TRUE`, excluding subgroups 15 and 23. In the call to the `qcc` function, the argument `D[inc]` is the vector containing the number of nonconforming cans for all the the subgroups in the original 30 samples minus groups 15 and 23. This is the set that will be used to calculate the control limits. The argument `sizes=size[inc]` is the vector of subgroup sizes associated with these 28 subgroups. The argument `newdata=D[!trial]` specifies that the data for nonconformities in the additional 24 subgroups in data frame `orangejuice`, where `trial=FALSE` should be plotted on the control chart with the previously calculated limits. The argument `newsizes=size[!trial])` is the vector of subgroup sizes associated with the 24 new subgroups. For more information and examples of selecting subsets of data frames (see Wickham and Grolemund[102]).

The immediate impression from the new chart is that the process, after adjustments, is operating at a new lower proportion defective. The run of consecutive points below the center line confirms that impression. The obvious assignable cause for this shift is successful adjustments.

Based on this improved performance, the control limits should be recalculated based on the new data. This was done, and additional data was collected over the next 5 shifts of operation (subgroups 55 to 94). The R code below first loads the data from subgroups 31 to 54, which is in the data frame orangejuice2 in the R package qcc. In this data set trial=TRUE for subgroups 31 to 54, and trial=FALSE for subgroups 55 to 94. The qcc function call in the code then plots all the data on the control chart with limits calculated from just subgroups 31 to 54. The result is shown in Figure 4.8.

```
R>library(qcc)
R>data(orangejuice2)
R>attach(orangejuice2)
R>names(D) <- sample
R>qcc(D[trial], sizes=size[trial], type="p",
  newdata=D[!trial], newsizes=size[!trial])
R>detach(orangejuice2)
```

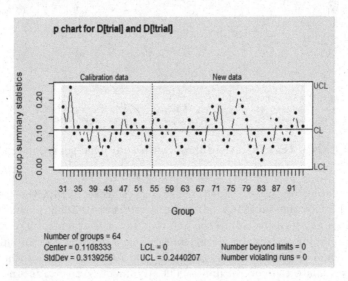

FIGURE 4.8: Revised limits with additional subgroups

In this chart there is no indication of assignable causes in the new data. The proportion nonconforming varies between 0.02 to 0.22 with an average proportion nonconforming of 0.1108. This is not an acceptable quality level. Due to the fact that no assignable causes are identified in Figure 4.7, the *p* chart alone gives no insight on how to improve the process and reduce the number of nonconforming items. This is one of the shortcomings of attribute charts in Phase I studies. Another shortcoming, pointed out by Borror, and Champ[8], is that the false alarm rate for *p* and *np* charts increases with the number of subgroups and could lead to wasted time trying to identify phantom assignable causes.

To try to improve the process to an acceptable level of nonconformities, the next step might be to use some of the seven tools to be described in Section 4.4. For example, classify the 351 nonconforming items found in subgroups 31 to 94 and display them in a Pareto Chart. Seeing the most prevalent types of nonconformities may motivate some ideas among those familiar with the process as to their cause. If there is other recorded information about the processing conditions during the time subgroups were collected, then scatter diagrams and correlation coefficients with the proportion nonconforming may also stimulate ideas about the cause for differences in the proportions. These ideas can be documented in cause-and-effect diagrams, and may lead to some Plan-Do-Check-Act investigations. More detail on these ideas will be shown in Section 4.4. Even more efficient methods of investigating ways to improve a process using the Experimental Designs are described in the Chapter 5.

Continual investigations of this type may lead to additional discoveries of ways to lower the proportion nonconforming. The control chart (like Figure 4.7) serves as part of a logbook to document the timing of interventions and their subsequent effects. Assignable causes discovered along the way are also documented in the OCAP for future use in Phase II.

The use of the control charts in this example, along with the other types of analysis suggested in the last two paragraphs, would all be classified as part of the Phase I study. Compared to the simple example from Mitra[70] presented in the last section, this example gives a clearer indication of what might be involved in a Phase I study. In Phase I, the control charts are not used to monitor the process but rather as tools to help identify ways to improve and to record and display progress.

Phase I would be complete when process conditions are found where the performance is stable (in control) at a level of nonconformities for attribute charts, or a level of process variability for variable charts that is acceptable to the customer or next process step. An additional outcome of the Phase I study is an estimate of the current process average and standard deviation.

4.3.2 Constructing other types of Attribute Charts with qcc

When the number of items inspected in each subgroup is constant, np charts for nonconformites are preferable. The actual number of nonconformites for each subgroup are plotted on the np chart rather than the fraction nonconforming. The control limits, which are based on the Binomial distribution, are calculated with the following formulas:

$$UCL = n\bar{p} + 3\sqrt{n\bar{p}(1-\bar{p})}$$

$$\text{Center line} = n\bar{p} \tag{4.2}$$

$$LCL = n\bar{p} - 3\sqrt{n\bar{p}(1-\bar{p})}$$

where, n is constant so that the control limits remain constant for all subgroups. The control limits for a p chart, on the other hand, can vary depending upon the subgroup size. The R code below illustrates the commands to produce an np chart with qcc using the data from Figure 14.2 in [19]. In this code, sizes=1000 is specified as a single constant rather than a vector containing the number of items for each subgroup.

```
R>library(qcc)
R>d<-c(9,12,13,12,11,9,7,0,12,8,9,7,11,10)
R>qcc(d,sizes=1000,type="np")
```

When charting nonconformities, rather than the number of nonconforming items in a subgroup, a c-chart is more appropriate. When counting nonconforming (or defective) items, each item in a subgroup is either conforming or nonconforming. Therefore, the number of nonconforming in a subgroup must be between zero and the size of the subgroup. On the other hand, there can be more than one nonconformity in a single inspection unit. In that case, the Poisson distribution is better for modeling the situation. The control limits for c-chart are based on the Poisson distribution and are given by the following formulas:

$$UCL = \bar{c} + 3\sqrt{\bar{c}}$$

$$\text{Center line} = \bar{c} \tag{4.3}$$

$$LCL = \bar{c} - 3\sqrt{\bar{c}}$$

where, $\bar{c}$ is the average number of defects per inspection unit.

Montgomery[72] describes an example where a c-chart was used to study the number of defects found in groups of 100 printed circuit boards. More than one defect could be found in a circuit board so that the upper limit for defects is greater than 100. This data is also included in the qcc package. When the inspection unit size is constant (i.e. 100 circuit boards) as it is in

this example, the `sizes=` argument in the call to qcc is unnecessary. The R
code below, from the qcc documentation, illustrates how to create a *c*-chart.

```
R>library(qcc)
R>data(circuit)
R>attach(circuit)
R>qcc(circuit$x[trial], sizes=circuit$size[trial], type="c")
```

u-charts are used for charting defects when the inspection unit does not
have a constant size. The control limits for the *u*-chart are given by the following formulas:

$$UCL = \overline{u} + 3\sqrt{\overline{u}/k}$$

$$\text{Center line} = \overline{u} \tag{4.4}$$

$$LCL = \overline{u} + 3\sqrt{\overline{u}/k}$$

where, $\overline{u}$ is the average number of defects per a standardized inspection unit
and k is the size of each individual inspection unit. The R code to create
an *u*-chart using the data in Figure 14.3 in [19] is shown below. The qcc()
function used in that code uses the formulas for the control limits shown in
Equation 4.4 and are not constant. [19] replaces k, in Equation 4.4, with $\overline{k}$,
the average inspection unit size. Therefore, the control limits in that book are
incorrectly shown constant. If you make the chart using the code below, you
will see that the control limits are not constant.

```
R>library(qcc)
R>d<-c(6,7,8,8,6,7,7,6,3,1,2,3,3,4)
R>qcc(d,sizes=k,type="u")
```

4.4 Finding Root Causes and Preventive Measures

4.4.1 Introduction

When an assignable cause appears on a control chart, especially one constructed with retrospective data, the reason for that assignable cause is not
always as obvious as the reasons for late bus arrival times on Patrick Nolans's
control chart shown in section 4.1. It may take some investigation to find the
reason for, and the way to eliminate, an assignable cause. The people most

qualified to investigate the assignable causes that appear on a control chart are those on the front line who actually work in the process that produced the out-of-control points. They have the most experience with the process, and their accumulated knowledge should be put to work in identifying the reasons for assignable causes. When people pool their ideas, spend time discovering faults in their work process, and are empowered to change the process and correct those faults, they become much more involved and invested in their work. In addition, they usually make more improvements than could ever be made by others less intimately involved in the day-to-day process operation.

However, as Deming pointed out, there were often barriers that prevented employee involvement. Often management was reluctant to give workers authority to make decisions and make changes to a process, even though they were the ones most involved in the process details. New workers often received inadequate training and no explanation of why they should follow management's rigid rules. Testing the employee's competency in correct process procedures was usually ignored, and management stressed production as the most important goal. If workers experienced issues, they were powerless to stop the process while management was slow to react to their concerns. Management only got involved when one of the following two things occurred: a production goal was not met, or a customer complained about some aspect of quality.

To address this problem, prior to 1950, Deming wrote what became point 12 in his 14 Points for Management. This point is as follows:

12. Remove barriers to the pride of workmanship. People are eager to do a good job and are distressed when they cannot. To often misguided supervisors, faulty equipment, and defective materials stand in the way. These barriers must be removed. (see Walton[98]).

Olmstead[76] remarked that assignable causes can usually be identified by the data patterns they produce. A list of these data patterns are: single points out of control limits, shift in the average or level, shift in the spread or variability, gradual change in process level (trend), and a regular change in process level (cycle).

To increase and improve the use of workers' process knowledge, Kaoru Ishikawa, a Deming Prize recipient in Japan, introduced the idea of quality circles. In quality circles, factory foremen and workers met together to learn problem-solving tools that would help in investigating ways to eliminate assignable causes and improve processes.

The problem-solving tools that Ishakawa emphasized were called the 7-tools. This list of tools has evolved over time, and it is not exactly the same in every published description. For purposes of this book they are:

1. Flowchart

2. Cause-and-Effect Diagram

3. Check-sheets or Quality Information Data

4. Pareto Diagrams

5. Scatter Plots

6. Defect Identification Reports

7. Control Charts

Phase I control charts have already been discussed in the last chapter. The remainder of this chapter will be devoted to discussing Shewhart Cycle (or PDCA) along with the first six of the seven tools. Emphasis will be placed on how these tools can be used to finding the reasons for and eliminating assignable causes.

4.4.2   Flowcharts

When multiple people (possibly on different shifts) work in the same process, they might not all understand the steps of the process in the same way. Out of control signals may be generated when someone performs a task in one process step differently than others. Process flowcharts are useful for helping everyone who works in a process to gain the same understanding of the process steps. The symbols typically used in a flowchart are shown in Figure 4.9

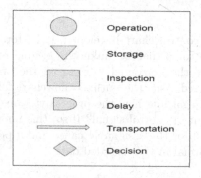

FIGURE 4.9: Flowchart Symbols

Process flowcharts are usually developed as a group exercise by a team of workers. The initial step in completing the flowchart is for the team to come to a consensus and construct the initial flowchart. Next, when everyone working in the process is following the steps as shown in the flowchart, the process results should be observed. If the process output meets the desired objective, the flowchart should be finalized and used as the reference. If the desired process output is not met, the team should reassemble and discuss their opinions regarding the reasons for poor performance and consider modifications to the flowchart. When consensus is reached on a modified flowchart the cycle begins again. This process of developing a flowchart is shown in Figure 4.10.

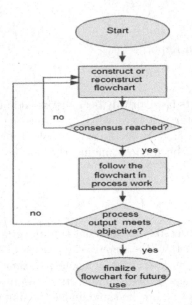

FIGURE 4.10: Creating a Flowchart

When a Phase I control chart developed with stored retrospective data shows an out-of-control signal of unknown origin, reference the process flowchart and involve the personnel working in the process when the out-of-control data occurred. Was the outlined flowchart followed at that time? Are there any notes regarding how the process was operated during and after the time of the out of control signal? If so, this information may lead to discovering the reason for the assignable cause and a possible remedy, if the process returns to normal in the recorded data.

4.4.3   PDCA or Shewhart Cycle

If one or more steps in a previously finalized process flowchart were not followed, it may or may not be a reason for an observed out-of-control signal. It could have been caused by some other change that was not noticed or recorded. If you want to discover the cause for an out-of-control signal, you need to do more than simply observe the process in action. An often quoted statement by George Box is "To find out what happens to a system (or process) when you interfere with it, you have to interfere with it not just passively observe it"[9]. This is the basis of the PDCA cycle first described by Shewhart and represented in Figure 4.11.

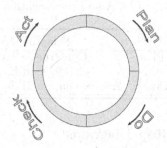

FIGURE 4.11: PDCA or Shewhart Cycle

The Shewhart Cycle will be demonstrated using flowchart example above.

1. Plan–in this step a plan would be made to investigate the potential cause of the out-of-control signal (i.e., the failure to follow one of the process steps as outlined on the flow diagram.)

2. Do–in this step the modified process step that was followed at the time of the out-of-control-signal will be purposely followed again at the present time.

3. Check–in this step the process output that is measured and recorded on a control chart will be checked to see if another out-of-control signal results after the purposeful change in procedure.

4. Act—if the out-of-control signal returns, it can be tentatively assumed to have been caused by the purposeful change made in step 2. These first three steps are often repeated (or replicated) several times, as indicated by the circular pattern in Figure 5.3, before proceeding to the Act step. When the results of the action performed in the Do step are consistent, it strengthens the conclusion and provides strong evidence for taking action in the Act step of the PDCA. The "Act" taken should then be to prevent the change that was made from occurring in the future. This action would be classified as a preventative action. In Japan during the 1960s Shigeo Shingo, an industrial engineer at Toyota, defined the term Poka-Yoke to mean mistake proofing. Using this method, Japanese factories installed safeguards, such as lane departure warning signals on modern automobiles, to prevent certain mistakes from occurring.

The first three steps of the PDCA or Shewhart cycle are similar to the steps in the scientific method as shown below:

TABLE 4.2: PDCA and the Scientific Method

Step	PDCA	Scientific Method
1	Plan	Construct a Hypothesis
2	Do	Conduct an Experiment
3	Check	Analyze data and draw Conclusions

4.4.4   Cause-and-Effect Diagrams

The plan step in the PDCA cycle can begin with any opinions or hypothesis concerning the cause for trouble. One way to record and display opinions or hypotheses is to use a cause-and-effect diagram. These are preferably made in a group setting where ideas are stimulated by interaction within a group of people. The opinions recorded on a cause-and-effect diagram are organized into major categories or stems off the main horizontal branch, as illustrated in the initial diagram shown in Figure 4.12

Another way to construct a cause-and-effect diagram is to use a process flow diagram as the main branch. In that way, the stems branch off the process steps where the hypotheses or leaves naturally reside.

One way to come up with the major categories or stems is through the use of an affinity diagram (see Chapter 8 Holtzblatt et. al.[40]). In the first step of constructing an affinity diagram, everyone in the group writes their ideas or opinions about the cause of a problem on a slip from a notepad or a sticky note. It is important that each person in the group is allowed to freely express their opinion, and no one should criticize or try to suppress ideas at this point. The notes are then put on a bulletin board or white board. Next, members of the group try to reorganize the notes into distinct categories of similar ideas by moving the location of the notes. Notes expressing the same idea can be combined at this point. Finally, the group decides on names for each of the categories and an initial cause-and-effect diagram like Figure 4.12 is constructed, with a stem off the main branch for each category.

The next step is to add leaves to the to the stems. The ideas for the leaves are stimulated by asking why. For example asking, "Why would road conditions cause a driver to lose control of the car?" This might result in the ideas or opinions shown on the right upper stem of the diagram shown in Figure 4.13 Leaves would be added to the other stems in the figure, by following the same process.

More detail can be found by asking why again at each of the leaves. For example asking, "Why would the tire blow out?" This might result in the opinion added to the blow out leaf on the upper left of Figure 4.14.

When the cause-and-effect diagram is completed, members of the group that made the cause-and-effect diagram can begin to test the ideas that are felt by consensus to be most important. The tests are performed using the PDCA cycle to verify or discredit the ideas on the cause-and-effect diagram.

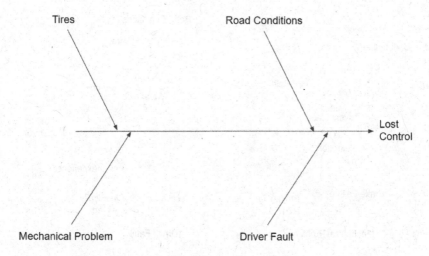

FIGURE 4.12: Cause-and-Effect-Diagram Major Categories

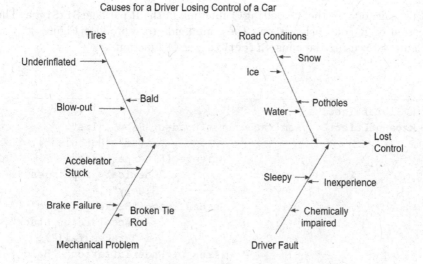

FIGURE 4.13: Cause-and-Effect-Diagram–First Level of Detail

Cause-and-effect diagrams, like flow diagrams, are usually handwritten. However, a printed cause-and-effect diagram showing the first level of detail (like Figure 4.13) can be made by the `causeEffectDiagram()` function in the

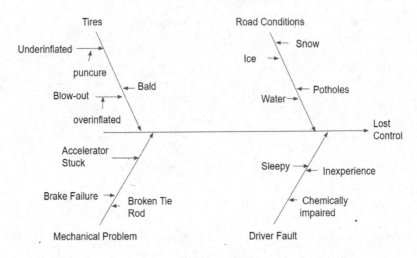

Causes for a Driver Losing Control of a Car

FIGURE 4.14: Cause-and-Effect-Diagram–Second Level of Detail

R package qcc, or the ss.ceDiag() function in the R package SixSigma. The section of R code below shows the commands to reproduce Figure 4.13 as Figure 4.15, using the causeEffectDiagram() function.

```
R>library(qcc)
R>causeEffectDiagram(cause=list(Road=c("Snow","Ice",
                                "Water","Potholes"),
                    Driver=c("Inexperience",
                             "Chemical _impairment",
                             "Sleepy"),
                    Mechanical=c("Broken_tie_rod",
                                 "Accelerator_stuck",
                                 "Brakes_fail"),
                    Tires=c("Underinflation","Bald",
                            "Overinflation")),
                    effect="Lost Control")
```

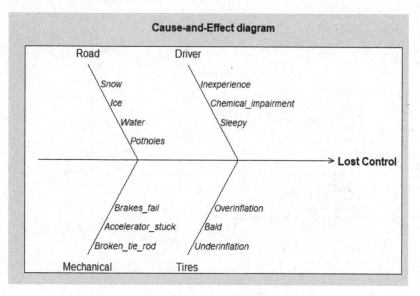

FIGURE 4.15: qcc Cause-and-Effect Diagram

4.4.5 Check Sheets or Quality Information System

Check sheets are used for collecting data. For example Figure 4.16 shows a Defective item check sheet that would be used by inspectors to record detailed information as they inspect items.

Ishikawa defined other types of check sheets such as:

1. Defect Location

2. Defective Cause

3. Check-up confirmation

4. Work Sampling

These were devised so that people involved in work within a process could quickly record information on paper to be filed for future reference. Having available data gives insight into the current level of performance, and can be a guide to future action. However, in today's world, having records stored on paper in file cabinets is obsolete technology and has been replaced by computer files.

A Quality Information System (QIS) is the modern equivalent of check sheets filed by date. A QIS contains quality-related records from customers, suppliers and internal processes. It consists of physical storage devices (the devices on which data is stored such as servers and cloud storage), a network that provides connectivity, and an infrastructure where the different sources

Check Sheet

Product: _____   Date: _____
                                  Location: _____

Manufacturing Stage: _____
Number inspected: _____   Inspector's name _____
Remarks: _____

                                  Lot No. _____
                                  Order No. _____

Type	Check	Subtotal
Contamination	//	2
Sealing Failure	//// //// //// //// //// //// ////	30
Adhesive Failure	////	5
Material Defects	/	1
Printing Off Color	//// ////	9
Ink Migration	///	3
	Grand Total:	50

FIGURE 4.16: Defective Item Check Sheet

of data are integrated. Most large companies maintain a QIS (like those the FDA requires companies in the pharmaceutical industry to maintain).

A QIS(see Burke and Silvestrini[14]) can be used to:

- Record critical data (e.g. measurement and tool calibration)

- Store quality process and procedures

- Create and deploy operating procedures (e.g. ISO 9001–based quality management system)

- Document required training

- Initiate action (e.g. generate a shop order from a customer order)

- Control processes (controlling the operation of a laser cutting machine within specification limits)

- Monitor processes (e.g. real time production machine interface with control charting)

- Schedule resource usage (e.g personnel assignments)

- Manage a knowledge base (e.g. capturing storing and retrieving needed knowledge)

- Archive data (e.g. customer order fulfillment)

The data in a QIS is often stored in relational databases and organized in a way so that portions of it can be easily be retrieved for special purposes. Additionally, routine reports or web pages called "dashboards" are produced periodically so managers can see current information about what has happened and what is currently happening. Quality control tools described in this book, like control charts and process capability studies, can provide additional insight about what is actually expected to happen in the near future. Phase I control charts and related process capability studies can be produced with the data retrieved from the QIS. In addition, when data retrieved from the QIS is used to determine the root cause of out-of-control signals, it can explain the cause for previous undesirable performance, and give direction for what can be done to prevent it in the future.

SQL (pronounced "ess-que-el") stands for Structured Query Language. SQL is used to communicate with a relational database. According to ANSI (American National Standards Institute), it is the standard language for relational database management systems. An open source relational database management system based on SQL (MySQL) is free to download and use. There is also an enterprise version that is available for purchase on a subscription basis. There are functions in R packages such as RODBC, RJDBC, and rsqlserver that can connect to an SQL data base and directly fetch portions of the data. An article giving examples of this is available at https://www.red-gate.com/simple-talk/sql/reporting-services/making-data-analytics-simpler-sql-server-and-r/.

4.4.6 Line Graphs or Run Charts

Whether an assignable cause seen on a control chart represents a data pattern, like a shift in the process average or variance or a trend or cyclical pattern, it is often helpful to determine if similar data patterns exist in other variables representing process settings or conditions that are stored in the QIS. A quick way to examine data patterns is to make a line graph or run chart.

A line graph or run chart is constructed by plotting the data points in a sequence on a graph with the y-axis scaled to the values of the data points, and the x-axis tic marks numbered 1 through the number of values in the sequence. For example, Figure 4.17 shows a line graph of a portion of the data from Figure 13 in O'Neill et. al.[77]. The data represent the standard deviations of plaque potency assays of a reference standard from monthly measurement studies.

It can be seen that the assay values decrease after month 7. If this change coincided with the beginning or end of an out-of-control signal on a retrospective control chart that was detected by a run of points above or below the center line, it could indicate that measurement error was the cause of the problem. Of course, further investigation and testing would be required for

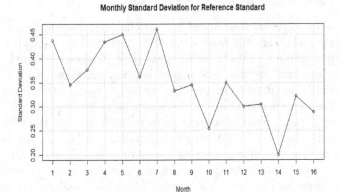

FIGURE 4.17: Standard Deviations of Plaque Potency Assays of a Reference Standard

verification. Line graphs can be made with the plot function in R as illustrated in the code below.

```
R>s<-c(.45,.345,.375,.435,.45,.36,.46,.335,.348,.252,.35,.30,
.302,.20,.325,.29)
R>plot(s,type="o",lab=c(17,5,4),ylab="Standard Deviation",
xlab="Month",main = "Monthly Standard Deviation for
Reference Standard")
R>grid()
```

In this code, s<-c(.45,.345,.375,.435,.45,.36,.46,.335, ...) are the monthly standard deviations, and the command `grid()` adds a grid to the graph with horizontal and vertical lines at the x and y axis tic marks.

4.4.7   Pareto Diagrams

To find the reason for a change in attribute data, such as an increase in the number for nonconforming items, it is useful to classify or divide the nonconforming items into categories. The Pareto diagram is a graphical tool that clearly exposes the relative magnitude of of various categories. It consists of a bar chart with one bar representing the total number in each category. The bars in the Pareto diagram are arranged by height in descending order with the largest on the left. The bars are augmented by a line graph of the cumulative percent in the combined categories from left to right. Pareto charts can be produced using the `pareto.chart()` function in the qcc package, the

ParetoChart() function in the qualityTools package, or the paretochart() function in the qicharts package. For example, if the defective item counts shown in Figure 4.16 represent a classification of the defective orange juice cans in subgroup 23 shown in Figure 4.6, then the Pareto diagram in Figure 4.18 can be constructed using the pareto.chart() function in the qcc package as shown in the code below.

```
R>library(qcc)
R>nonconformities<-c(2,30,5,1,9,3)
R>names(nonconformities)<-c("Contamination","Sealing Failure",
   "Adhesive Failure","Material Defects","Printing Off Color",
   "Ink Migration")
R>paretoChart(nonconformities, ylab = "Nonconformance
   frequency",main="Pareto Chart")
```

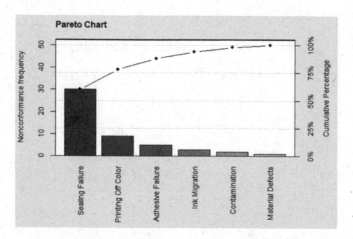

FIGURE 4.18: Pareto Chart of Noconforming Cans Sample 23

The scale on the left of the graph is for the count in each category and the scale on the right is for the cumulative percent. This Pareto diagram clearly shows that most of defective or nonconforming cans in subgroup 23 were sealing failures that accounted for over 50% of the total. Sealing failure and adhesive failure usually reflect operator error. This sparked an investigation of what operator was on duty at that time. It was found that the half-hour period when subgroup 23 was obtained occurred when a temporary and inexperienced operator was assigned to the machine; and that could account for the high fraction nonconforming.

4.4.8 Scatter Plots

Scatter plots are graphical tools that can be used to illustrate the relationship or correlation between two variables. A scatter plot is constructed by plotting an ordered pair of variable values on a graph where the x-axis usually represents a process variable that can be manipulated, and the y-axis is usually a measured variable that is to be predicted. The scatter plot is made by placing each of the ordered pairs on a graph where the x_i coordinate lines up with its value on the horizontal scale, and the y_i coordinate lines up with its value on the vertical scale.

Figure 4.19 is an example scatter plot patterened after one in Figure 2 in the paper by Cunningham and Shanthikumar[20]. The R code to produce the Figure is shown below,

```
R>plot(file$cycle_time, file$yield,xlab=c("Cycle
   Time"),ylab=c("Yield"),type='p',pch=20)
```

where `file$cycle_time` and `file$yield` are columns in an R data frame `file` retrieved from a database.

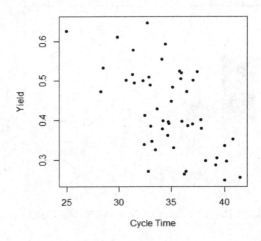

FIGURE 4.19: Scatter Plot of Cycle Time versus Yield

The x-variable is the cycle time in a semiconductor wafer fabrication facility, and the the y-variable is the yield. The plot shows a negative relationship where as the cycle time increases the process yield tends to decrease. Scatter plots can show negative, positive, or curvilinear relationships between two variables, or no relationship if the points on the graph are in a random pattern. Correlation does not necessarily imply causation and the increase in the x-variable and corresponding decrease in the y-variable may both have been caused by something different.

In the case of Figure 4.19, efforts had been made to reduce cycle times, since it is beneficial in many ways. Customer demands can be met more quickly, costs are reduced when there is less work-in-progress inventory, and the opportunity to enter new markets becomes available. If an assignable cause appeared on a control chart showing a reduction in contamination defects, while these efforts to reduce cycle time were ongoing, the reason for this reduction would need to be determined. If the reason were found, whatever was done to cause the reduction could be incorporated into the standard operating procedures. One way to look for a possible cause of the fortunate change is by examining scatter plots versus manipulated process variables. Figure 4.20 shows a scatter plot of cycle time versus contamination defects on dies within semiconductor wafers.

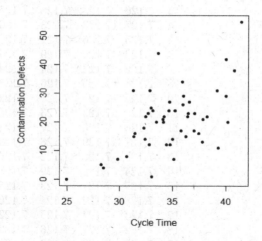

FIGURE 4.20: Scatter Plot of Cycle Time versus Contamination Defects

This plot shows a positive relationship. The plot itself did not prove that shortened cycle times decreased contamination defects, but it was reasoned that each wafer lot was exposed to possible contamination throughout processing. Therefore, it makes sense that reducing the cycle time reduces exposure and should result in fewer contamination defects and higher yields. When efforts continued to reduce cycle time, and contamination defects continued to decrease and yields continued to increase, these realizations provided proof that reducing cycle time had a positive effect on yields.

4.5   Process Capability Analysis

At the completion of a Phase I study, where control charts have shown that
the process is in control at what appears to be an acceptable level, then calculating a process capability ratio (PCR) is an appropriate way to document
the current state of the process.

For example, $\overline{X} - R$ charts were made using the data in Table 4.3 taken
from Christensen, Betz and Stein[19]

TABLE 4.3: Groove Inside Diameter Lathe Operation

Subgroup No.		Measurement		
1	7.124	7.122	7.125	7.125
2	7.123	7.125	7.125	7.128
3	7.126	7.128	7.128	7.125
4	7.127	7.123	7.123	7.126
5	7.125	7.126	7.121	7.122
6	7.123	7.129	7.129	7.124
7	7.122	7.122	7.124	7.124
8	7.128	7.125	7.126	7.123
9	7.125	7.125	7.121	7.122
10	7.126	7.123	7.123	7.125
11	7.126	7.126	7.127	7.128
12	7.127	7.129	7.128	7.129
13	7.128	7.123	7.122	7.124
14	7.124	7.125	7.129	7.127
15	7.127	7.127	7.123	7.125
16	7.128	7.122	7.124	7.126
17	7.123	7.124	7.125	7.122
18	7.122	7.121	7.126	7.123
19	7.128	7.129	7.122	7.129
20	7.125	7.125	7.124	7.122
21	7.125	7.121	7.125	7.128
22	7.121	7.126	7.120	7.123
23	7.123	7.123	7.123	7.123
24	7.128	7.121	7.126	7.127
25	7.129	7.127	7.127	7.124

When variables control charts were used in Phase I, then the PCR's C_p
and C_{pk} are appropriate. These indices can be calculated with the formulas:

$$C_p = \frac{USL - LSL}{6\sigma} \tag{4.5}$$

$$C_{pk} = \frac{Min(C_{Pu}, C_{Pl})}{3\sigma}, \tag{4.6}$$

where

$$C_{Pu} = \frac{USL - \overline{X}}{\sigma}$$

$$C_{Pl} = \frac{\overline{X} - LSL}{\sigma} \tag{4.7}$$

The measure of the standard deviation σ used in these formulas is estimated from the subgroup data in Phase I control charts. If $\overline{X} - R$ charts were used, then $\sigma = \overline{R}/d_2$, where d_2 is given in the vignette SixSigma::ShewhartConstants in the R package SixSigma. They can also be found in Section 6.2.3.1 of the online NIST Engineering Statistics Handbook. If $\overline{X} - s$ charts were used, then $\sigma = \overline{s}/c_4$, where c_4 is found in the same tables for the appropriate subgroup size n.

Assuming the data in Table 4.3 shows the process is in control, then the processCapability() function in the R package qcc, the cp() function in the R package qualityTools, or the ss.ca.study() function in the R package SixSigma can compute the capability indices and display the histogram of the data with the specification limits superimposed as in Figure 4.21. The code to do this is shown using the processCapability() function in the R package qcc. The output is below the code.

```
R>Lathe<-read.table("Lathe.csv",header=TRUE,sep=","
  ,na.strings="NA",dec=".",strip.white=TRUE)
R>library(qcc)
R>qcc(Lathe,type="R",plot=FALSE)
R>pc<-qcc(Lathe,type="xbar",plot=FALSE)
R>processCapability(pc,spec.limits=c(7.115,7.135))

-- Process Capability Analysis --------------------
Number of obs = 100          Target = 7.125
Center      = 7.1249         LSL    = 7.115
StdDev      = 0.002098106    USL    = 7.135
Capability indices  Value  2.5%  97.5%
             Cp     1.59   1.37   1.81
             Cp_l   1.57   1.38   1.76
             Cp_u   1.60   1.41   1.80
             Cp_k   1.57   1.34   1.80
             Cpm    1.59   1.37   1.81
```

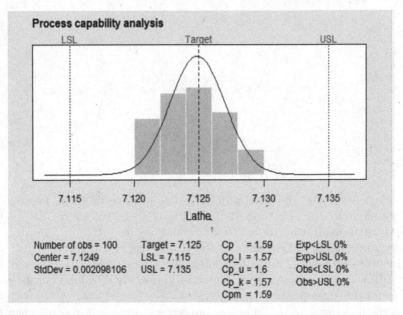

FIGURE 4.21: Capability Analysis of Lathe Data in Table 4.3

```
Exp<LSL 0%   Obs<LSL 0%
Exp>USL 0%   Obs>USL 0%
```

If the process is exactly centered between the lower and upper specification limits, then $C_p = C_{pk}$. In the example below, the center line $\overline{\overline{X}} = 7.1249$ is very close to the center of the specification range and therefore C_p and C_{pk} are nearly equal. When the process is not centered, C_{pk} is a more appropriate measure than C_p. The indices $C_{p_l} = 1.573$ and $C_{p_u} = 1.605$ are called Z_L and Z_U in some textbooks, and they would be appropriate if there was only a lower or upper specification limit. Otherwise $C_{pk} = min(C_{p_l}, C_{p_u})$. Since the capability indices are estimated from data, the report also gives 95% confidence intervals for them.

To better understand the meaning of the capability index, a portion of the table presented by Montgomery[72] is shown in Table 4.4. The second column of the table shows the process fallout in ppm when the mean is such that C_{p_l} or C_{p_u} is equal to the PCR shown in the first column. The third column shows the process fallout in ppm when the process is centered between the two spec limits so that C_P is equal to the PCR. The fourth and fifth columns show the process fallout in ppm if the process mean shifted left or right by 1.5σ after the PCR had been established.

In this table it can be seen that a $C_{p_l} = 1.50$ or $C_{p_u} = 1.50$ (when there is only a lower or upper specification limit) would result in only 4 ppm (or a proportion of 0.000004) out of specifications. If $C_p = 1.50$ and the process is

TABLE 4.4: Capability Indices and Process Fallout in PPM

| | Process Fallout (in ppm) | | Process Fallout with 1.5σ shift | |
PCR	Single Spec	Double Specs	Single Spec	Double Specs
0.50	66,807	133,614	500,000	501,350
1.00	1,350	2700	66,807	66,811
1.50	4	7	1,350	1,350
2.00	0.0009	0.0018	4	7

centered between upper and lower specification limits, then only 7 ppm (or a proportion of 0.000007) would be out of the specification limits. As shown in the right two columns of the table, even if the process mean shifted 1.5σ to the left or right after C_p was shown to be 1.50, there would still be only 1,350 ppm (or 0.00134 proportion) nonconforming. This would normally be an acceptable quality level (AQL).

Table 4.4 makes it clear why Ford motor company mandated that their suppliers demonstrate their processes were in a state of statistical control and had a capability index of 1.5 or greater. That way Ford could be guaranteed an acceptable quality level (AQL) for incoming components from their suppliers, without the need for acceptance sampling.

The validity of the PCR C_p is dependent on the following.

1. The quality characteristic measured has a normal distribution.

2. The process is in a state of control.

3. For two sided specification limits, the process is centered.

The control chart and output of the **process.capability()** function serve to check these. The histogram of the Lathe data in Figure 4.21 appears to justify the normality assumption. Normal probability plots would also be useful for this purpose.

When Attribute control charts are used in a Phase I study, C_p and C_{pk} are unnecessary. When using a p or np chart, the final estimate $\bar{p}$ is a direct estimate of the process fallout. For c and u charts, the average count represents the number of defects that can be expected for an inspection unit.

4.6 OC and ARL Characteristics of Shewhart Control Charts

The operating characteristic OC for a control chart (sometimes called β) is the probability that the next subgroup statistic will fall between the control limits and not signal the presence of an assignable cause. The OC can be plotted versus the hypothetical process mean for the next subgroup, similar to the OC curves for sampling plans discussed in Chapters 2 and 3.

4.6.1   OC and ARL for Variables Charts

The function `occurves()` in the R package `qcc` can make OC curves for $\overline{X}$-control charts created with `qcc`. In this case

$$OC = \beta = P(LCL \leq \overline{X} \leq UCL). \tag{4.8}$$

For example, the R code below demonstrates using the `occurves` function to create the OC curve based on the revised control chart for the Coil resistance data discussed in Section 4.2.1.

```
R>Coilm <- read.table("Coil.csv",  header=TRUE,sep=",",
  na.strings="NA",dec=".",strip.white=TRUE)
R>library(qcc)
R>beta <- ocCurves(pc, nsigmas=3)
```

Figure 4.22 shows the resulting OC curves for future $\overline{X}$ control charts for monitoring the process with subgroups sizes of 5, 1, 10, 15, and 20 (the first number is the subgroup size for the control chart stored in `pc`). It can be seen that if the process shift for the next subgroup is small (i.e., less that .5 σ), then the probability of falling within the control limits is very high. If the process shift is larger, the probability of falling within the control limits (and not detecting the assignable cause) decreases rapidly as the size of the subgroup increases.

Figure 4.23 gives a more detailed view. In this figure, it can be seen that when the subgroup size is $n = 5$ and the process mean shifts to the left or right by 1 standard deviation, there is about an 80% chance that the next mean plotted will fall within the control limits and the assignable cause will not be detected. On the other hand, if the process mean shifts by 2 standard deviations, there is less that a 10% chance that it will not be detected. On the right side of the figure, it can be seen that the chance of not detecting a 1 or 2 standard deviation shift in the mean is much higher when the subgroup size is $n = 1$.

If the mean shifts by $k\sigma$ just before the ith subgroup mean is plotted on the $\overline{X}$ chart, then the probability that the first mean that falls out of the control limits is for the $(i + m)$th subgroup is:

$$[\beta(\mu + k\sigma)]^{m-1}(1 - \beta(\mu + k\sigma)). \tag{4.9}$$

This is a Geometric probability with parameter $\beta(\mu + k\sigma)$. Therefore, the expected number of subgroups before a mean shift is discovered, or the average run length (ARL), is given by

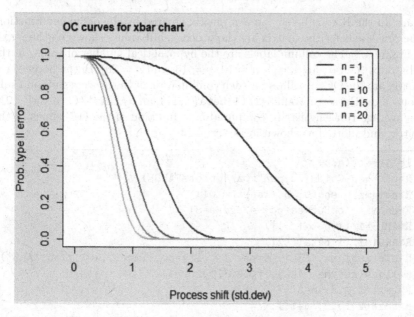

FIGURE 4.22: OC curves for Coil Data

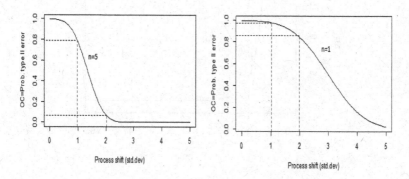

FIGURE 4.23: OC curves for Coil Data

$$ARL = \frac{1}{(1 - \beta(\mu + k\sigma))}, \tag{4.10}$$

the expected value of the Geometric random variable.

Although the qcc package does not have a function to plot the ARL curve for a control chart, the ocCurves function can store the OC curves for the $\overline{X}$

chart. In the R code below, `pc` is a matrix created by the `occurves` function. The row labels for the matrix are the process shift values shown on the x-axis of Figure 4.8. The column labels are the hypothetical number of values in the subgroups (i.e., $n = 5$, $n = 1$, $n = 10$, $n = 15$, and $n = 20$). In the body of the matrix are the OC=β values for each combination of the process shift and subgroup size. By setting `ARL5=1/(1-beta[ ,1])` and `ARL1=1/(1-beta[ ,2])` the average run lengths are computed as a function of the OC values in the matrix and plotted as shown in Figure 4.24.

```
R>library(qcc)
R>pc<-qcc(Coilm, type="xbar",plot=FALSE)
R>nsigma = seq(0, 5, length=101)
R>beta <- ocCurves(pc, nsigmas=3)
R>ARL5=1/(1-beta[ ,1])
R>ARL1=1/(1-beta[ ,2])
R>plot(nsigma, ARL1, type='l',,lty=3,ylim=c(0,20),ylab='ARL')
R>lines(nsigma, ARL5, type='l',lty=1)
R>text(1.2,5,'n=5')
R>text(2.1,10,'n=1')
```

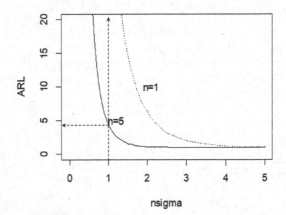

FIGURE 4.24: ARL curves for Coil Data with Hypothetical Subgroup Sizes of 1 and 5

In this figure, it can be seen that if the subgroup size is $n = 5$, and the process mean shifts by 1 standard deviation, it will take 5 (rounded up to an integer) subgroups on the average before before the $\overline{X}$ chart shows an out of control signal. If the subgroup size is reduced to $n = 1$, it will take many more subgroups on the average (over 40) before an out of control signal is displayed.

4.6.2 OC and ARL for Attribute Charts

The ocCurves function in the qcc package can also make and plot OC curves for p and c type attribute control charts. For these charts, β is computed using the Binomial or Poisson distribution. For example, the R code below makes the OC curve for the p chart shown in Figure 4.6 for the orange juice can data. The result is shown in Figure 4.25.

```
R>data(orangejuice)
R>beta <- with(orangejuice,occurves(qcc(D[trial],
  sizes=size[trial], type="p", plot=FALSE)))
R>print(round(beta, digits=4))
```

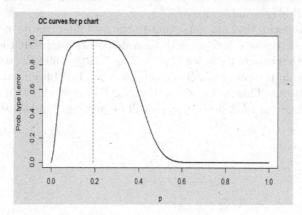

FIGURE 4.25: OC curves for Orange Juice Can Data, Subgroup Size=50

The center line is $\bar{p} = 0.231$ in Figure 4.6, but in Figure 4.25 it can be seen that the probability of the next subgroup of 50 falling within the control limits of the p chart is ≥ 0.90 for any value of p between 0.15 and about 0.30. So the p chart is not very sensitive to detecting small changes in the proportion nonconforming.

ARL curves can be made for the p chart in a way similar to the last example by copying the OC values from the matrix beta created by the ocCurves function. OC curves for c charts can be made by changing type="p" to type="c".

4.7 Summary

Shewhart control charts can be used when the data is collected in rational subgroups. The pooled estimate of the variance within subgroups becomes the estimated process variance and is used to calculate the control limits. This calculation assumes the subgroup means are normally distributed. However, due to the Central Limit Theorem, averages tend to be normally distributed and this is a reasonable assumption.

If the subgroups were formed so that assignable causes occur within the subgroups, then the control limits for the control charts will be wide and few if any assignable cause will be detected. If a control chart is able to detect assignable causes, it indicates that the variability within the subgroups is due mainly common causes.

The cause for out of control points are generally easier discover when using Shewhart variable control charts (like $\overline{X}-R$ or $\overline{X}-s$) than Shewhart attribute charts (like p-charts, np-charts, c-charts, or u-charts). This is important in Phase I studies where one of the purposes is to establish an OCAP containing a list of possible assignable causes.

If characteristics that are important to the customer can be measured and variables control charts can be used, then demonstrating that the process is in control with a capability index $C_p \geq 1.50$ or $C_{pk} \geq 1.50$ will guarantee that production from the process will contain less that 1,350 ppm nonconforming or defective items. This is the reason Deming[21] asserted that no inspection by the customer (or next process step) will be necessary in this situation.

4.8 Exercises

1. Run all the code examples in the chapter, as you read.

2. Recreate the Cause and Effect Diagram shown below using the `causeEffectDiagram()` function or the `ss.ceDiag()` function, and add any additional ideas you have to the diagram.

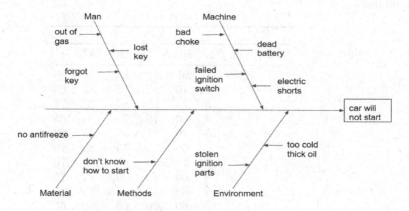

FIGURE 4.26: CE Diagram for exercise

3. Assuming that you believe that a dead battery is the most likely cause for the car not starting, write one or more sentences describing what you would do at each of the PDCA steps to solve this problem.

4. Construct a Flowchart showing the operations, inspection points, and decision points in preparing for an exam over chapters 1–4.

5. Using the data in the R code block in Section 4.2.5, use R to construct a line graph (like Figure 4.1) of Patrick Nolan's data.

6. Create an OC curve and an ARL curve for the $\overline{X}$-chart you create using the data from Table 4.3.

7. Use the R code with the `qcc` function at the beginning of Section 4.3 to plot the p chart. Why do the control limits vary from subgroup to subgroup?

8. If variables control charts (such as $\overline{X} - R$ or $\overline{X} - s$) were used and the process was shown to be in control with a $C_{pk} = 1.00$, then

   (a) What proportion nonconforming or out of the specification limits would be expected? (i.e., show how the values in Table 4.4 were obtained)

(b) If an attribute p-chart were used instead of the $\overline{X} - R$ or $\overline{X} - s$ charts, then what subgroup size would be necessary to show the same in control proportion nonconforming as the variables charts in (a) with $C_{pk} = 1.00$?

9. Consider the data in Table 4.5 taken from the Automotive Industry Action Committee [96].

TABLE 4.5: Phase I Study for Bend Clip Dim "A" (Specification Limits .50 to .90 mm)

Subgroup No.	Bend Clip Dim "A"				
1	.65	.70	.65	.65	.85
2	.75	.85	.75	.85	.65
3	.75	.80	.80	.70	.75
4	.60	.70	.70	.75	.65
5	.70	.75	.65	.85	.80
6	.60	.75	.75	.85	.70
7	.75	.80	.65	.75	.70
8	.60	.70	.80	.75	.75
9	.65	.80	.85	.85	.75
10	.60	.70	.60	.80	.65
11	.80	.75	.90	.50	.80
12	.85	.75	.85	.65	.70
13	.70	.70	.75	.75	.70
14	.65	.70	.85	.75	.60
15	.90	.80	.80	.75	.85
16	.75	.80	.75	.80	.65
17	.75	.70	.85	.70	.80
18	.75	.70	.60	.70	.60
19	.65	.65	.85	.65	.70
20	.60	.60	.65	.60	.65
21	.50	.55	.65	.80	.80
22	.60	.80	.65	.65	.75
23	.80	.65	.75	.65	.65
24	.65	.60	.65	.60	.70
25	.65	.70	.70	.60	.65

(a) Make an R-chart for the data. If any points are out of the control limits on the R-chart, assume that assignable causes were discovered and added to the OCAP.

(b) Remove the subgroups of data you found out of control on the R chart, and construct $\overline{X}$ and R charts with the remaining data.

(c) Check for out of control signals on the $\overline{X}$ chart (use the Western Electric Rules in addition to checking points out of the control limits).

(d) Assume that assignable causes were discovered for any out of control points found in subgroups 18 to 25 only (since it was discovered that out of spec raw materials were used while making those subgroups).

(e) Eliminate any out-of-control subgroups you discovered between 18 to 25, and then make $\overline{X}$ and R charts with the remaining data.

(f) If there are no out of control signals in the control charts you created, then calculate the appropriate capability index and a graph similar to Figure 4.21.

(g) Is the process capable of meeting customer requirements? If not, what is the first thing you might suggest to improve it.

(h) If defective bend clips were classified by the cause resulting in the table below, use R to construct a Pareto diagram.

Cause	Number of occurrences
Out of spec raw material	121
Part misplaced on die	15
Bend allowance	9
Spring back	47

5

DoE for Troubleshooting and Improvement

5.1 Introduction

All work is performed within a system of interdependent processes. Components of processes include things such as people, machines, methods, materials, and the environment. This is true whether the work being performed provides a product through a manufacturing system, or a service.

Variation causes a system or process to provide something less than the desired result. The variation is introduced into the process by the components of the process as mentioned above. The variation from the ideal may be of two forms. One form is inconsistency or too much variation around the ideal output, and another is askew or off-target performance. Deming has often been quoted as having said "improving quality is all about reducing variability".

The variability of process output can be classified into the two categories of: (1) assignable or special causes, and (2) common causes, as described in Chapter 4. The methods for effectively reducing common cause variability are much different than the methods for reducing variability due to assignable causes. The differences in the methods appropriate for reducing variability of each type are described next.

If process outputs are measured and plotted on control charts, assignable causes are identified by out-of-control signals/indexout-of-control signals. Usually the cause for an out-of-control signal on a Phase I Control Chart is obvious. Because out-of-control signals occur so rarely when a process is stable, examination of all the circumstances surrounding the specific time that an out-of-control signal occurred should lead to the discovery of unusual processing conditions which could have caused the problem. Tools such as the 5 W's (*who, what, where, when,* and *why*), and ask why 5 times are useful in this respect.

Once the cause is discovered, a corrective action or adjustment should immediately be made to prevent this assignable cause from occurring again in the future. The efficacy of any proposed corrective action should be verified using a PDCA. This is a four step process. In the first step (plan) a proposed corrective action is described. In the second step (do) the proposed plan is carried out. In the third step (check) the effects of the corrective action are observed. If the action has corrected the problem proceed to the fourth step (act). In this step the corrective action is institutionalized by adding it to the

OCAP described in Chapter 4, or implementing a preventative measure. If the corrective action taken in the third step did not correct the problem return to the first step (plan) and start again. The PDCA is often referred to as the Shewhart Cycle (who originally proposed it) or the Deming Cycle, since he strongly advocated it.

Common causes of variability cannot be recognized on a control chart, which would exhibit a stable pattern with points falling within the control limits with no unusual trends or patterns. Common causes don't occur sporadically. They are always present. To discover common causes and create a plan to reduce their influence on process outputs requires careful consideration of the entire process, not just the circumstances immediately surrounding an out of control point.

Deming used his red bead experiment, and the funnel experiment to demonstrate the folly of blaming common cause problems on the immediate circumstances or people and then making quick adjustments. When the cause of less than ideal process outputs is due to common causes, quick adjustments are counter productive and could alienate the workforce and make them fearful to cooperate in discovering and reducing the actual causes of problems.

The way to reduce common cause variability is to study the entire process. In today's environment of "Big Data" and Data Science, data analytics could be proposed as a way of studying the entire process and discovering ways to improve it. Through data analytics, past process information and outputs recorded in a quality information system (Burke and Silvestrini[14]) would be mined to discover relationships between process outputs and processing conditions. By doing this, it is felt that standardizing on the processing conditions that correlate with better process performance will improve the process.

However, this approach is rarely effective. The changes in process conditions to affect an improvement were usually unknown and not recorded in databases. Further, correlations found studying recorded data often do not represent cause and effect relationships. These correlations could be the result of cause and effect relations with other unrecorded variables.

A better way to study the entire process and find ways to improve is through the use of a SIPOC (Britz et. al.[11]) diagram (**S**uppliers provide **I**nputs that the **P**rocess translates to **O**utputs for the **C**ustomer) as shown in Figure 5.1. Here we can see that the output of a process is intended to satisfy or please the customer.

To discover changes necessary to counteract the influence of common causes and improve a process requires process knowledge and or intuition. When people knowledgeable about the process consider the inputs, processing steps, methods, and machines that translate the inputs to outputs, they can usually hypothesize potential reasons for excessive variation in outputs or off target performance. A list of potentially beneficial changes to the process can be gathered in brainstorming sessions. Affinity diagrams, cause-and-effect

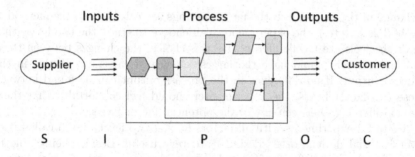

FIGURE 5.1: SIPOC Diagram.

diagrams, FMEA, and Fault Tree Analysis are useful ways for documenting ideas and opinions expressed in these sessions.

Once a list is developed, the potential changes can be tested during process operation. Again, the quote of George Box[9] is appropriate, "To find out what happens to a system (or process) when you interfere with it, you have to interfere with it not just passively observe it". In this context, using data analytics, or attempting to discover potential process improvements by analyzing process variables stored in databases is what George Box means by passively observing a process. This method may not be successful because the changes needed to improve a process may have been unknown and not recorded in the databases. For this reason, correlations discovered in these exercises may not imply cause and effect relationships, and changes to the process based on these observed correlations might not lead to the anticipated process improvements.

To discover cause-and-effect relationships, changes must be made and the results of the changes must be observed. The PDCA approach is one way to test potential changes to a process. It is similar to the Edisonian technique of holding everything constant and varying one factor at a time to see if it causes a change to the process output of interest. This approach is popular in many disciplines, but it is not effective if changes in the process output are actually caused by simultaneous changes to more than one process factor. In addition, if a long list of potential changes to process factors to improve a process have been developed in a brainstorming session, it will be time consuming to test each one separately using the PDCA approach.

One method of simultaneously testing changes to several factors, and identifying joint (or interaction) effects between two or more factors is through use of a factorial experimental design. In a factorial design, all possible combinations of factor settings are tested. For example, if three process variables were each to be studied at four levels, there would be a total of $4 \times 4 \times 4 = 4^3 = 64$ combinations tested. This seems like a lot of testing, but the number can usually be reduced by restricting the number of levels of each factor to two.

If the factor can be varied over a continuous range (like temperature), then the experimenter should choose two levels as far apart as reasonable to evoke a change in the response. If the factor settings are categorical (like method 1, method 2, ... etc.), the experimenter can choose to study the two levels that he or she feels are most different. By this choice, the changes between these two factor levels would have the largest chance of causing a change in the process output. If no difference in the process output can be found between these two factor levels, the experimenter should feel comfortable that there are no differences between any of the potential factor levels.

Factorial experiments with all factors having two levels (often called two-level factorial designs or 2^k designs) are very useful. Besides being able to detect simultaneous effects of several factors, they are simple, easy to set up, easy analyze, and their results are easy to communicate. These designs are very useful for identifying ways to improve a process. If a Phase I control chart shows that the process is stable and in control, but the capability index (PCR) is such that the customer need is not met, then experimentation with the process variables or redesigning the process must be done to find a solution. Use of a 2^k experimental design is an excellent way to do this. These designs are also useful for trying to identifying the cause or remedy for an out of control situation that was not immediately obvious.

2^k designs require at least 2^k experiments, and when there are many potential factors being studied, the basic two-level factorial designs still require a large number of experiments. This number can be drastically reduced using just a fraction of the total runs required for a 2^k design, and fractional factorial, 2^{k-p} designs, are again very useful for identifying ways to improve a process.

ISO (The International Standards Organization) Technical Committee 69 recognized the importance of 2^k, and 2^{k-p} experimental designs and response surface designs for Six Sigma and other process improvement specialists, and they issued technical reports ISO/TC 29901 in 2007 and ISO/TR 12845 in 2010 (Johnson and Boulanger[44]). These technical reports were guideline documents. ISO/TC 29901 offered a template for implementation of 2^k experimental designs for process improvement studies, and ISO/TR 12845 a template for implementation of 2^{k-p} experimental designs for process improvement studies.

Although 2^k experiments are easy to set up, analyze, and interpret, they require too many experiments if there are 5 or 6 factors or more. While 2^{k-p} designs require many fewer experiments, the analysis is not always so straight forward, especially for designs with many factors. Other experimental design plans such as the alternative screening sesigns in 16 runs[47], and the Definitive Screening Designs [48], can handle a large number of factors yet offer reduced run size and straightforward methods of analysis, using functions in the R package daewr. The Definitive Screening designs allow three-level factors and provide the ability to explore curvilinear response surface type models that the ISO Technical Committee 69 also recognized as important for process improvement studies.

After some unifying definitions, this chapter will discuss 2^k designs and fractions of 2^k designs, alternative screening designs, and Definitive Screening designs and provide examples of creating and analyzing data from all of these using R.

5.2 Definitions:

1. **An Experiment (or run)** is conducted when the experimenter changes at least one of the factor levels he or she is studying (and has control over) and then observes the result, i.e., do something and observe the result. This is not the same as passive collection of observational data.

2. **Experimental Unit** is the item under study upon which something is changed. This could be human subjects, raw materials, or just a point in time during the operation of a process.

3. **Treatment Factor (or independent variable)** is one of the variables under study that is being controlled near some target value, or *level*, during any experiment. The level is being changed in some systematic way from run to run in order to determine the effect the change has on the response.

4. **Lurking Variable (or Background variable)** is a variable that the experimenter is unaware of or cannot control, but it does have an influence on the response. It could be an inherent characteristic of the experimental units (which are not identical) or differences in the way the treatment level is applied to the experimental unit, or the way the response is measured. In a well-planned experimental design the effect of these lurking variables should balance out so as to not alter the conclusions.

5. **Response (or Dependent Variable)** is the characteristic of the experimental units that is measured at the completion of each experiment or run. The value of the response variable depends on the settings of the independent variables (or factor levels) and the lurking variables.

6. **Factor Effect** is the difference in the expected value or average of all potential responses at the high level of a factor and the expected value or average of all potential responses at the low level of a factor. It represents the expected change in the response caused by changing the factor from its low to high level. Although this effect is unknown, it can be estimated as the difference in sample average responses calculated with data from experiments.

7. **Interaction Effect** If there is an interaction effect between two factors, then the effect of one factor is different depending on the level of the other

factor. Other names for an interaction effect are joint effect or simultaneous effect. If there is an interaction between three factors (called a three-way interaction), then the effect of one factor will be different depending on the combination of levels of the other two factors. Four-factor interactions and higher order interactions are similarly defined.

8. **Replicate runs** are two or more experiments with the same settings or levels of the treatment factors, but using different experimental units. The measured response for replicate runs may be different due to changes in lurking variables or the inherent characteristics of the experimental units.

9. **Duplicates** refers to duplicate measurements of the same experimental unit from one run or experiment. Differences in the measured response for duplicates is due to measurement error and these values should be averaged and not treated as replicate runs in the analysis of data.

10. **Experimental Design** is a collection of experiments or runs that is planned in advance of the actual execution. The particular experiments chosen to be part of an experimental design will depend on the purpose of the design.

11. **Randomization** is the act of randomizing the order that experiments in the experimental design are completed. If the experiments are run in an order that is convenient to the experimenter rather than in a random order, it is possible that changes in the response that appear to be due to changes in one or more factors may actually be due to changes in unknown and unrecorded lurking variables. When the order of experimentation is randomized, this is much less likely to occur due to the fact that changes in factors occur in a random order that is unlikely to correlate with changes in lurking variables.

12. **Confounded Factors** results when changes in the level of one treatment factor in an experimental design correspond exactly to changes in another treatment factor in the design. When this occurs, it is impossible to tell which of the confounded factors caused a difference in the response.

13. **Biased Factor** results when changes in the level of one treatment factor, in an experimental design, correspond exactly to changes in the level of a lurking variable. When a factor is biased, it is impossible to tell if any resulting changes in the response that occur between runs or experiments is due to changes in the biased factor or changes in the lurking variable.

14. **Experimental error** is the difference in observed response for one particular experiment and the long run average response for all potential experimental units that could be tested at the same factor settings or levels.

5.3 2^k Designs

2^k designs have k factors each studied at 2 levels. Figure 5.2 shows an example of a 2^3 design (i.e., $k = 3$). There are a total of $2 \times 2 \times 2 = 8$ runs in this design. They represent all possible combinations of the low and high levels for each factor. In this figure the factor names are represented as A, B, and C, and the low and high levels of each factor are represented as $-$ and $+$, respectively. The list of all runs is shown on the left side of the figure where the first factor (A) alternates between low and high levels for each successive pair of runs. The second factor (B) alternates between pairs of low and high levels for each successive group of four runs, and finally the last factor (C) alternates between groups of four consecutive low levels and four consecutive high levels for each successive group of eight runs. This is called the standard order of the experiments.

The values of the response are represented symbolically as y with subscripts representing the run in the design. The right side of the figure shows that each run in the design can be represented graphically as the corners of a cube. If there are replicate runs at each of the 8 combination of factor levels, the responses (y) in the figure could be replaced by the sample averages of the replicates $(\bar{y})$ at each factor level combination.

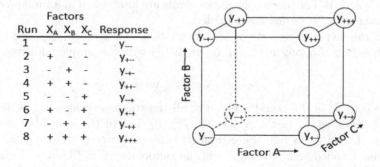

FIGURE 5.2: Symbolic representation of a 2^3 Design.

With the graphical representation in Figure 5.2, it is easy to visualize the estimated factor effects. For example, the effect of factor A is the difference in the average of the responses on the right side of the cube (where factor A is at its high level) and the average of the responses on the left side of the cube. This is illustrated in Figure 5.3.

The estimated effect of factor B can be visualized similarly as the difference in the average of the responses at the top of the cube and the average of the responses at the bottom of the cube. The estimated effect of factor C is the

	Factors			
Run	X_A	X_B	X_C	Response
1	-	-	-	y_{---}
2	+	-	-	y_{+--}
3	-	+	-	y_{-+-}
4	+	+	-	y_{++-}
5	-	-	+	y_{--+}
6	+	-	+	y_{+-+}
7	-	+	+	y_{-++}
8	+	+	+	y_{+++}

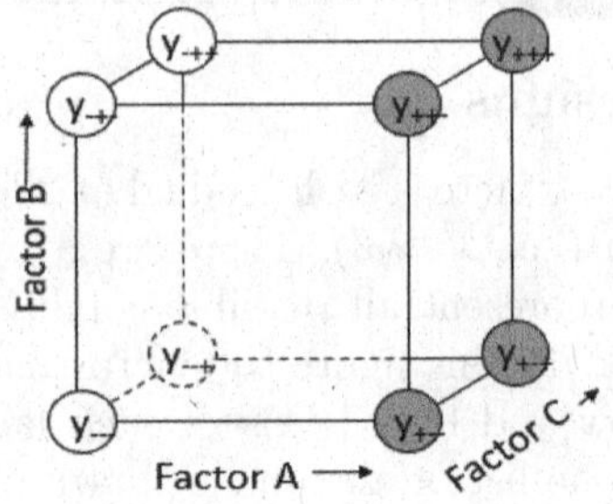

Effect of Factor A=E_A =$(y_{+--} + y_{++-} + y_{+-+} + y_{+++})/4 - (y_{---} + y_{-+-} + y_{--+} + y_{-++})/4$

FIGURE 5.3: Representation of the Effect of Factor A.

difference in the average of the responses on the back of the cube and the average of the responses at the front of the cube.

In practice the experiments in a 2^k design should not be run in the standard order, as shown in Figure 5.1. If that were done, any changes in a lurking variable that occurred near midway through the experimentation would bias factor C. Any lurking variable that oscillated between two levels could bias factors A or B. For this reason experiments are always run in a random order, once the list of experiments is made.

If the factor effects are independent, then the average response at any combination of factor levels can be predicted with the simple additive model

$$y = \beta_0 + \beta_A X_A + \beta_B X_B + \beta_C X_C. \tag{5.1}$$

Where β_0 is the grand average of all the response values; β_A is half the estimated effect of factor A (i.e., $E_A/2$); β_B is half the estimated effect of factor B; and β_C is half the estimated effect of factor C. X_A, X_B, and X_C are the $\pm$ coded and scaled levels of the factors shown in Figures 5.1 and 5.2. The conversion between actual factor levels and the coded and scaled factor levels are accomplished with a coding and scaling equation like:

$$X_A = \frac{\text{Factor level} - \left(\frac{\text{High level+Low level}}{2}\right)}{\left(\frac{\text{High level–Low level}}{2}\right)}. \tag{5.2}$$

This coding and scaling equation converts the high factor level to +1, the low factor level to −1, and factor levels between the high and low to a value between −1 and +1. When a factor has qualitative levels like: Type A, or Type B. One level is designated as +1 and the other as −1. In this case predictions from the model can only be made at these two levels of the factor.

If the factor effects are dependent, then effect of one factor is different at each level of another factor or each combination of levels of other factors. For example, the estimated conditional main effect of factor B, calculated when factor C is held constant at its low level is

$$(E_{B|C=-} = (y_{-+-} + y_{++-})/2 - (y_{---} + y_{+--})/2).$$

This can be visualized as the difference in the average of the two responses in the gray circles at the top front of the cube on the left side of Figure 5.4 and the average of the two responses in the gray circles at the bottom front of the cube shown on the left side of Figure 5.4. The estimated conditional main effect of factor B when factor C is held constant at its high level is

$$(E_{B|C=+} = (y_{-++} + y_{+++})/2 - (y_{--+} + y_{+-+})/2).$$

It can be visualized on the right side of Figure 5.4 as the difference of the average of the two responses in the gray circles at top back of the cube and the average of the two responses in the gray circles at the back bottom of the cube.

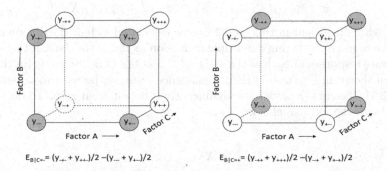

FIGURE 5.4: Conditional Main Effects of B given the level of C

The B×C interaction effect is defined as half the difference in the two estimated conditional main effects as shown in Equation 5.3.

$$
\begin{aligned}
E_{BC} &= E_{B|C=+} - E_{B|C=-} \\
&= (y_{-++} + y_{+++} + y_{---} + y_{+--})/4) \\
&\quad - (y_{-+-} + y_{++-} + y_{--+} + y_{+-+})/4
\end{aligned}
\tag{5.3}
$$

Again this interaction effect can be seen to be the difference in the average of the responses in the gray circles and the average of the responses in the white circles shown in Figure 5.5.

If the interaction effect is not zero, then additive model in Equation 5.1 will not be accurate. Adding half of the estimated interaction effect to model

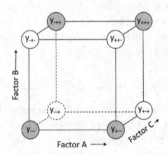

FIGURE 5.5: BC interaction effect

5.1 results in model 5.4 which will have improved predictions if the effect of factor B is different depending on the level of Factor C, or the effect of Factor C is different depending on the level of factor B.

$$y = \beta_0 + \beta_A X_A + \beta_B X_B + \beta_C X_C + \beta_{BC} X_B X_C \qquad (5.4)$$

Each β coefficient in this model can be calculated as half the difference in two averages, or by fitting a multiple regression model to the data. If there are replicate responses at each of the eight combinations of factor levels (in the 2^3 design shown in Figure 5.1) then significance tests can be used to determine which coefficients (i.e., βs) are significantly different from zero. To do that, start with the full model 5.4:

$$\begin{aligned} y =& \beta_0 + \beta_A X_A + \beta_B X_B + \beta_C X_C \\ &+ \beta_{AB} X_A X_B + \beta_{AC} X_A X_C + \beta_{BC} X_B X_C + \beta_{ABC} X_A X_B X_C. \end{aligned} \qquad (5.5)$$

The full model can be extended to 2^k experiments as shown in Equation 5.6.

$$y = \beta_0 + \sum_{i=1}^{k} \beta_i X_i + \sum_{i=1}^{k} \sum_{j \neq i}^{k} \beta_{ij} X_i X_j + \cdots + \beta_{i\ldots k} X_i \cdots X_k. \qquad (5.6)$$

Any insignificant coefficients found in the full model can be removed to reach the final model.

5.3.1 Examples

In this section two examples of 2^k experiments will used to illustrate the R code to design, analyze, and interpret the results.

5.3.2   Example 1: A 2^3 Factorial in Battery Assembly

A 2^3 experiment was used to study the process of assembling nickel-cadmium batteries. This example is taken from Ellis Ott's book *Process Quality Control-Troubleshooting and Interpretation of Data* [78].

Nickel-Cadmium batteries were assembled with the goal of having a consistently high capacitance. However much difficulty was encountered in during the assembly process resulting in excessive capacitance variability. A team was organized to find methods to improve the process. This is an example of a common cause problem. Too much variability all of the time.

While studying the entire process as normally would be done to remove a common cause problem, the team found that two different production lines in the factory were used for assembling the batteries. One of these production lines used a different concentration of nitrate than the other line. They also found that two different assembly lines were used in the factory. One line was using a shim in assembling the batteries and the other line did not. In addition, two processing stations were used in the factory; at one station, fresh hydroxide was used, while reused hydroxide was used in the second station. The team conjectured that these differences in the assembly could be the cause of the extreme variation in the battery capacitance. They decided it would be easy to set up a 2^3 experiment varying the factor levels shown in Table 5.1, since they were already in use in the factory.

TABLE 5.1:   Factors and Levels in Nickel-Cadmium Battery Assembly Experiment

Factors	Level $(-)$	Level $(+)$
A–Production line	low level of nitrate	high level of nitrate
B–Assembly line	using shim in assembly	no shim used in assembly
C–Processing Station	using fresh hydroxide	using reused hydroxide

The team realized that if they found differences in capacitance caused by the two different levels of factor A, then the differences may have been caused by either differences in the level of nitrite used, or other differences in the two production lines, or both. If they found a significant effect of factor A, then further experimentation would be necessary to isolate the specific cause. A similar situation existed for both factors B and C. A significant effect of factor B could be due to whether a shim was used in assembly or due to other differences in the two assembly lines. Finally, a significant effect of factor C could be due to differences in the two processing stations or caused by the difference in the fresh and reused hydroxide.

Table 5.2 shows the eight combinations of coded factor levels and the responses for six replicate experiments or runs conducted at each combination. The capacitance values were coded by subtracting the same constant from each value (which will not affect the results of the analysis). The actual experiments were conducted in a random order so that any lurking variables like properties of the raw materials would not bias any of the estimated factor effects or interactions. Table 5.2 is set up in the standard order of experiments that was shown in Figure 5.1.

TABLE 5.2: Factor Combinations and Response Data for Battery Assembly Experiment

X_A	X_B	X_C			Measured Ohms			
−	−	−	−0.1	1.0	0.6	−0.1	−1.4	0.5
+	−	−	0.6	0.8	0.7	2.0	0.7	0.7
−	+	−	0.6	1.0	0.8	1.5	1.3	1.1
+	+	−	1.8	2.1	2.2	1.9	2.6	2.8
−	−	+	1.1	0.5	0.1	0.7	1.3	1.0
+	−	+	1.9	0.7	2.3	1.9	1.0	2.1
−	+	+	0.7	−0.1	1.7	1.2	1.1	−0.7
+	+	+	2.1	2.3	1.9	2.2	1.8	2.5

The design can be constructed in standard order to match the published data using the `fac.design()` and `add.response()` functions in the R package DoE.Base as shown in the section of code below.

```
R>library(DoE.base)
R>design<-fac.design(nlevels=c(2,2,2),replications=6,
  randomize=F,factor.names=list(A=c("nitrate-1",
  "nitrate-2"),B=c("Shim","No Shim"),
  C=c("Fresh","Reused")))
R>Capacitance<-c( -.1,  .6,  .6, 1.8, 1.1, 1.9,  .7, 2.1,
              1.0,  .8, 1.0, 2.1,  .5,  .7, -.1, 2.3,
               .6,  .7,  .8, 2.2,  .1, 2.3, 1.7, 1.9,
              -.1, 2.0, 1.5, 1.9,  .7, 1.9, 1.2, 2.2,
             -1.4,  .7, 1.3, 2.6, 1.3, 1.0, 1.1, 1.8,
               .5,  .7, 1.1, 2.8, 1.0, 2.1, -.7, 2.5)
R>add.response(design,Capacitance)
```

The R function `lm()` can be used to fit the full model (Equation 5.4) by least-squares regression analysis to the data from the experiment using the code below. Part of the output is shown below the code.

```
R>mod1<-lm(Capacitance~A*B*C, data=design)
R>summary(mod1)


Coefficients:
            Estimate Std. Error t value Pr(>|t|)
(Intercept)  1.18750    0.08434  14.080  < 2e-16 ***
A1           0.54583    0.08434   6.472 1.03e-07 ***
B1           0.32917    0.08434   3.903 0.000356 ***
C1           0.11667    0.08434   1.383 0.174237
A1:B1        0.12083    0.08434   1.433 0.159704
A1:C1        0.04167    0.08434   0.494 0.623976
B1:C1       -0.24167    0.08434  -2.865 0.006607 **
A1:B1:C1     0.03333    0.08434   0.395 0.694768
Signif. codes:  0 '***' 0.001 '**' 0.01 '*' 0.05 '.' 0.1 ' ' 1

Residual standard error: 0.5843 on 40 degrees of freedom
Multiple R-squared:  0.6354,Adjusted R-squared:  0.5715
F-statistic: 9.957 on 7 and 40 DF,  p-value: 3.946e-07
'''
```

In the output, it can be seen that factor A and B are significant as well as the two-factor interaction BC. There is a positive regression coefficient for factor A. Since this coefficient is half the effect of A=(production line or level of nitrate) it means that a higher average capacitance is expected using production line 2 where the high concentration of nitrate was used.

Because there is a significant interaction of factors B and C, the main effect of factor B cannot be interpreted in the same way as the main effect for factor A. The significant interaction means the conditional effect of factor B=(Assembly line or use of a shim) is different depending on the level of factor C=(Processing station and fresh or reused hydroxide). The best way to visualize the interaction is by using an interaction plot. The R function `interaction.plot()` is used in the section of code below to produce the interaction plot shown in Figure 5.6.

```
R>Assembly<-design$B
R>Hydroxide<-design$C
R>interaction.plot(Assembly,Hydroxide,Capacitance,type="b",
  pch=c(1,2),col=c(1,1))
```

In this figure it can be seen that the effect of using a shim in assembly reduces the capacitance by a greater amount when fresh hydroxide at processing station 1 is used than when reused hydroxide at processing station two is used.

However, the highest expected capacitance of the assembled batteries is predicted to occur when using fresh hydroxide and assembly line 1 where no shim was used.

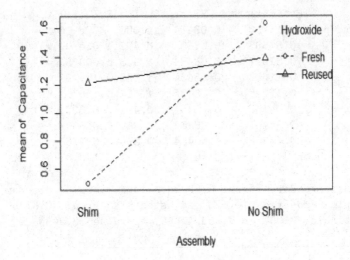

FIGURE 5.6: BC Interaction Plot

The fact that there was an interaction between assembly line and processing station, was at first puzzling to the team. This caused them to further investigate and resulted in the discovery of specific differences in the two assembly lines and processing stations. It was conjectured that standardizing the two assembly lines and processing stations would improve the battery quality by making the capacitance consistently high. A pilot run was made to check, and the results confirmed the conclusions of the 2^3 design experiment.

The assumptions required for a least-squares regression fit are that the variability in the model residuals (i.e., actual minus model predictions) should be constant across the range of predicted values, and that the residuals should be normally distributed. Four diagnostic plots for checking these assumptions can be easily made using the code below, and are shown in Figure 5.7.

```
par(mfrow=c(2,2))
plot(mod1)
par(mfrow=c(1,1))
```

The plot on the left indicates the spread in the residuals is approximately equal for each of the predicted values. If the spread in the residuals increased noticeably as the fitted values increased it would indicate that a more accurate model could be obtained by transforming the response variable before using the lm() function (see Lawson[57]) for examples. The plot on the right is a normal probability plot of the residuals. Since the points fall basically along the diagonal straight line, it indicates the normality assumption is satisfied. To understand how far from the straight line the points may lie before indicating the normality assumption is contradicted, you can make repeated normal probability plots of randomly generated data using the commands:

```
R>z<-rnorm(40,0,1)
R>qqnorm(z)
```

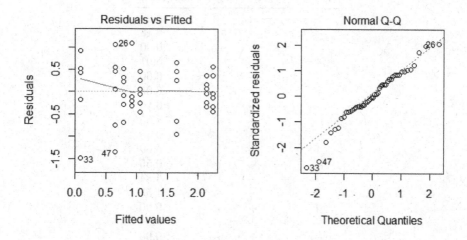

FIGURE 5.7: Model Diagnostic Plots

5.3.3 Example 2: Unreplicated 2^4 Factorial in Injection Molding

The second example from Durakovic[24] illustrates the use of a 2^4 factorial, with only one replicate of each of the 16 treatment combinations. The purpose of the experiments was to improve the quality of injection-molded parts by reducing the excessive flash. This is again a common cause that makes the average flash excessive (off target). The factors that were under study and their low and high levels are shown in Table 5.3.

A full factorial design and the observed flash size for each run are shown in Table 5.4. This data is in standard order (first column changing fastest etc.) but the experiments were again run in random order.

In the R code below, the FrF2() function in the R package FrF2() is used to create this design. Like the design in Example 1, it was created in

TABLE 5.3: Factors and Levels for Injection Molding Experiment

Factors	Level (−)	Level (+)
A–Pack Pressure in Bar	10	30
B–Pack Time in sec.	1	5
C–Injection Speed in mm/sec	12	50
D–Screw Speed in rpm	100	200

TABLE 5.4: Factor Combinations and Response Data for Injection Molding Experiment

X_A	X_B	X_C	X_D	Flash(mm)
−	−	−	−	0.22
+	−	−	−	6.18
−	+	−	−	0.00
+	+	−	−	5.91
−	−	+	−	6.60
+	−	+	−	6.05
−	+	+	−	6.76
+	+	+	−	8.65
−	−	−	+	0.46
+	−	−	+	5.06
−	+	−	+	0.55
+	+	−	+	4.84
−	−	+	+	11.55
+	−	+	+	9.90
−	+	+	+	9.90
+	+	+	+	9.90

standard order using the argument `randomize=F` in the `FrF2` function call. This was to match the already collected data shown in Table 5.4. When using the `fac.design()` or `FrF2()` function to create a data collection form prior to running experiments, the argument `randomize=F` should be left out and the default will create a list of experiments in random order. The random order will minimize the chance of biasing any estimated factor effect with the effect of any lurking variable that may change during the course of running the experiments.

```
R>library(FrF2)
R>design2<-FrF2(16,4,factor.names=list(A=c(10,30),B=c(1,5),
       C=c(12,50),D=c(100,200)),randomize=F)
R>Flash<-c(.22,6.18,0,5.91,6.6,6.05,6.76,8.65,0.46,
       5.06,0.55,4.84,11.55,9.9,9.9,9.9)
R>add.response(design2,Flash)
```

The R function lm() can be used to calculate the estimated regression coefficients for the full model (Equation 5.6 with $k = 4$). However, the Std. Error, t value, and Pr(>|t|) shown in the output for the first example cannot be calculated because there were no replicates in this example.

Graphical methods can be used to determine which effects or coefficients are significantly different from zero. If all the main effects and interactions in the full model were actually zero, then all of the estimated effects or coefficients would only differ from zero by an approximately normally distributed random error (due to the central limit theorem). If a normal probability plot were made of all the estimated effects or coefficients, any truly nonzero effects should appear as outliers on the probability plot.

The R code below shows the use of the R function lm() to fit the full model to the data, and a portion of the output is shown below the code.

```
R>mod2<-lm(Flash~A*B*C*D, data=design2)
R>summary(mod2)

Coefficients:
            Estimate Std. Error t value Pr(>|t|)
(Intercept)  5.783125         NA      NA       NA
A1           1.278125         NA      NA       NA
B1           0.030625         NA      NA       NA
C1           2.880625         NA      NA       NA
D1           0.736875         NA      NA       NA
A1:B1        0.233125         NA      NA       NA
A1:C1       -1.316875         NA      NA       NA
B1:C1        0.108125         NA      NA       NA
A1:D1       -0.373125         NA      NA       NA
B1:D1       -0.253125         NA      NA       NA
C1:D1        0.911875         NA      NA       NA
A1:B1:C1     0.278125         NA      NA       NA
A1:B1:D1    -0.065625         NA      NA       NA
A1:C1:D1    -0.000625         NA      NA       NA
B1:C1:D1    -0.298125         NA      NA       NA
A1:B1:C1:D1 -0.033125         NA      NA       NA

Residual standard error: NaN 'on 0 degrees of freedom
Multiple R-squared:       1,Adjusted R-squared:      NaN
F-statistic:   NaN on 15 and 0 DF, p-value: NA
```

The function fullnormal() from the daewr package can be used to make a normal probability plot of the coefficients (Figure 5.8) as shown in the code below.

```
R>library(daewr)
R>fullnormal(coef(mod2)[-1],alpha=.10)
```

The argument `coef(mod2)[-1]` requests a normal probability plot of the 15 coefficients in `mod2` (excluding the first term which is the intercept).

Normal Q-Q Plot

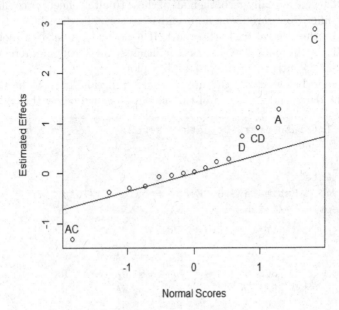

FIGURE 5.8: Normal Plot of Coefficients

In this plot it can be seen that main effects A=Pack Pressure in Bar, C=Injection Speed in mm/sec, and D=Screw Speed in rpm all appear to be significant since they do not fall along the straight line of insignificant effects. All three of these significant factors have positive coefficients which, would lead one to believe that setting all three of these factors to their low levels would minimize flash and improve the quality of the injected molded parts.

However, the CD interaction and the AC interaction also appear to be significant. Therefore, these two interaction plots should be studied before drawing any final conclusions. The R code to make these two interaction plots (Figures 5.9 and 5.10) is shown below.

```
R>A_Pack_Pressure<-design2$A
R>C_Injection_Speed<-design2$C
```

```
R>D_Screw_Speed<-design2$D
R>interaction.plot(A_Pack_Pressure,C_Injection_Speed,
        Flash,type="b",pch=c(1,2),col=c(1,2))
R>interaction.plot(C_Injection_Speed,D_Screw_Speed,
        Flash,type="b",pch=c(1,2),col=c(1,2))
```

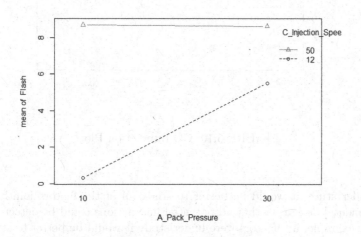

FIGURE 5.9: AC Interaction Plot

In Figure 5.9 it can be seen that factor A: Pack pressure only has a positive effect when factor C: Injection speed is set to its low level (12mm/sec). In Figure 5.10 it can be seen that the effect of Factor C: Injection speed is much stronger when Factor D: Screw speed is set to its high value of 200 rpm. Therefore the final recommendation was to run the injection molding machine at the low levels of Factor A (Pack Pressure=10 Bar), the low level of Factor C (Injection Speed=12 mm/sec) and the high level of Factor D (Screw Speed=200 rpm).

Factor B (Pack Time) was not significant nor was there any significant interaction involving Factor B. Therefore, this factor can be set to the level that is least expensive.

In the article by Durakovic[24], the reduced model involving main effects A, C, D and the AC and CD interactions was fit to the data and the normal plot of residuals and residuals by predicted values showed that the least squares assumptions were satisfied and that the predictions from this model would be accurate.

In both of the examples shown in this section, discovering significant interactions was key to solving the problem. In the real world, factor effects are rarely independent, and interactions are common. Therefore, if many factors

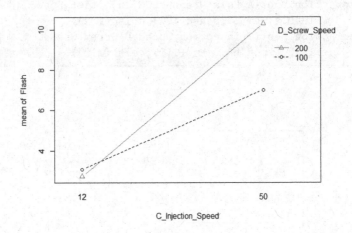

FIGURE 5.10: CD Interaction Plot

are under study it would be better to study all of the factors jointly in an experimental design, so that all potential interactions could be detected.

For example, if 6 factors were under study it would be better to study all factors simultaneously in a 2^6 factorial experiment than it would to study the first three factors in a 2^3 factorial experiment and the remaining three factors in another 2^3 factorial experiment. When running two separate 2^3 designs it would be impossible to detect interactions between any of the factors studied in the first experiment and any of the factors studied in the second experiment.

After brainstorming sessions where cause-and-effect diagrams are produced, there are often many potential factors to be investigated. However, the number of experiments required for a 2^k factorial experiment is 2^k. This can be large, for example 64 experiments are required for a 2^6 factorial and 1024 experiments required for a 2^{10} factorial.

Fortunately, many of the advantages obtained by running a 2^k full factorial can be obtained by running just a fraction of the total 2^k experiments. The next section will illustrate how to create a fraction of a 2^k design and examples will be used to show how the results can be interpreted.

5.4 2^{k-p} Fractional Factorial Designs

5.4.1 One-Half Fraction Designs

When creating a fraction of a 2^k factorial, the runs or experiments chosen to be in the fraction must be selected strategically. Table 5.5 shows what happens if the runs in a one-half fraction are not selected strategically. The left side of the table shows a 2^4 design in standard order. The column labeled "k-e" is an indicator of whether the run should be kept or eliminated from the fraction of the 2^4 design. The runs to be eliminated were chosen by a random selection process. The right side of the table shows the eight runs remaining in the one-half fraction after eliminating the runs with a "e" in the "k-e" column.

TABLE 5.5: Fractional Factorial by Random Elimination

k-e	run	X_A	X_B	X_C	X_D	run	X_A	X_B	X_C	X_D
k	1	$-$	$-$	$-$	$-$	1	$-$	$-$	$-$	$-$
e	2	$+$	$-$	$-$	$-$	4	$+$	$+$	$-$	$-$
e	3	$-$	$+$	$-$	$-$	5	$-$	$-$	$+$	$-$
k	4	$+$	$+$	$-$	$-$	8	$+$	$+$	$+$	$-$
k	5	$-$	$-$	$+$	$-$	10	$+$	$-$	$-$	$+$
e	6	$+$	$-$	$+$	$-$	12	$+$	$+$	$-$	$+$
e	7	$-$	$+$	$+$	$-$	13	$-$	$-$	$+$	$+$
k	8	$+$	$+$	$+$	$-$	16	$+$	$+$	$+$	$+$
e	9	$-$	$-$	$-$	$+$					
k	10	$+$	$-$	$-$	$+$					
e	11	$-$	$+$	$-$	$+$					
k	12	$+$	$+$	$-$	$+$					
k	13	$-$	$-$	$+$	$+$					
e	14	$+$	$-$	$+$	$+$					
e	15	$-$	$+$	$+$	$+$					
k	16	$+$	$+$	$+$	$+$					

The full 2^4 design on the left side of the table is orthogonal. This means that the runs where factor A is set at its low level, will have an equal number of high and low settings for factors B, C, and D. Likewise, the runs where factor A is set to its high level will have an equal number of high and low settings of factors B, C, and D. Therefore any difference in the average response between runs at the high and low levels of factor A cannot be attributed to factors B, C, or D because their effects would average out. Therefore, the effect of A is independent or uncorrelated with the effects of B, C, and D. This will be true for any pair of main effects or interactions when using the full 2^4 design.

On the other hand, when looking at the eight runs in the half-fraction of the 2^4 design shown on the right side of Table 5.5 a different story emerges. It can be seen that factor B is at its low level for every run where factor A is at

TABLE 5.6: Fractional Factorial by Confounding E and ABCD

run	X_A	X_B	X_C	X_D	$E = ABCD$
1	−	−	−	−	+
2	+	−	−	−	−
3	−	+	−	−	−
4	+	+	−	−	+
5	−	−	+	−	−
6	+	−	+	−	+
7	−	+	+	−	+
8	+	+	+	−	−
9	−	−	−	+	−
10	+	−	−	+	+
11	−	+	−	+	+
12	+	+	−	+	−
13	−	−	+	+	+
14	+	−	+	+	−
15	−	+	+	+	−
16	+	+	+	+	+

its low level, and factor B is at its high level for 4 of the 5 runs where factor A is at its high level. Therefore, if there is a difference in the average responses at the low and high level of factor A, it could be caused by the changes to factor B. In this fractional design, the effects of A and B are correlated. In fact, all pairs of effects are correlated although not to the high degree that factors A and B are correlated.

A fractional factorial design can be created that avoids the correlation of main effects using the *hierarchical ordering principle*. This principle, which has been established through long experience, states that main effects and low order interactions (like two-factor interactions) are more likely to be significant than higher order interactions (like 4 or 5-way interactions etc.). Recall that a two-factor interaction means that the effect of one factor is different depending on the level of the other factor involved in the interaction. Similarly, a four-factor interaction means that the effect of one factor will be different at each combination of levels of the other three factors involved in the interaction.

Higher order interactions are possible, but more rare. Based on that fact, a half-fraction design can be created by purposely confounding a main effect with a higher order interaction that is not likely to be significant. Table 5.6 illustrates this by creating a one-half fraction of a 2^5 design in 16 runs. This was done by defining the levels of the fifth factor, E, to be equal to the levels of the four way interaction ABCD. The settings for ABCD were created by multiplying the signs in columns A, B, C, and D together. For example for run number 1, ABCD=+, since the product of $(-) \times (-) \times (-) \times (-) = +$; likewise for run number 2 ABCD=− since $(+) \times (-) \times (-) \times (-) = -$.

In general, to create a one-half fraction of a 2^k design, start with the base design which is a full factorial in $k - 1$ factors (i.e., 2^{k-1} factorial). Next, assign the kth factor to the highest order interaction in the 2^{k-1} design.

If running a full 2^4 design using the factor settings in the first four columns in Table 5.6, the last column of signs ($ABCD$ in Table 5.6) would be used to calculate the four-way interaction effect. The responses opposite a $-$ sign in this $ABCD$ column would be averaged and subtracted from the average of the responses opposite a $+$ to calculate the effect.

In the half-fraction of a 2^5 design created by assigning the settings for the fifth factor, E, to the column of signs for $ABCD$, the four-way interaction would be assumed to be negligible, and no effect would be calculated for it. Instead, this column is used to both define the settings for a fifth factor E while running the experiments, and to calculate the effect of factor E once the data is collected from the experiments.

When eliminating half the experiments to create a one-half fraction, we cannot expect to estimate all the effects independently as we would in a full factorial design. In fact, only one-half of the effects can be estimated from a one-half fraction design. Each effect that is estimated is confounded with one other effect that must be assumed to be negligible. In the example of a $\frac{1}{2}2^5 = 2^{5-1}$ design in Table 5.6, main effect E is completely confounded with the $ABCD$ interaction effect. We cannot estimate both, and we rely on the hierarchical ordering principle to assume the four-way interaction is negligible.

To see the other effects that are are confounded we will define the *generator* and *defining relation* for the design. Consider E to represent the entire column of $+$ or $-$ signs in the design matrix in Table 5.6. Then the assignment $E = ABCD$ is called the *generator* of the design. It makes the column of signs for E, in the design matrix, identical to the column of signs for $ABCD$. Multiplying by E on both sides of the generator equation, $E = ABCD$, results in the equation $E^2 = ABCDE$, or $I = ABCDE$, where I is a column of $+$ signs that results from squaring each sign in the column for E. The equation $I = ABCDE$ is then called the *defining relation* of the design.

The defining relation can be used to find all the pairs of effects that are completely confounded in the half fraction design. The I on the left side of the defining relation equation is a column of $+$ signs. I is called the multiplicative identity because if it is multiplied (element-wise) by any column of signs in the design matrix the result will be the same (i.e. $A \times I = A$). If both sides of the defining relation are multiplied by a column in the design matrix, two confounded effects will be revealed, (i.e., $A \times (I = ABCDE) = (A = A^2 \times BCDE) = (A = BCDE)$). Therefore main effect A is confounded with the four-way interaction $BCDE$.

Multiplying both sides of the defining relation by each main effect and interaction that can be estimated from the base design (i.e., 2^{k-1}) results in the what is called the *alias pattern* that shows all pairs of confounded effects in the 2^{k-1} half fraction design. Equation 5.7 shows the alias pattern for the half-fraction design shown in Table 5.6.

$$A = BCDE$$
$$B = ACDE$$
$$C = ABDE$$
$$D = ABCE$$
$$E = ABCD$$
$$AB = CDE$$
$$AC = BDE$$
$$AD = BCE \tag{5.7}$$
$$AE = BCD$$
$$BC = ADE$$
$$BD = ACE$$
$$BE = ACD$$
$$CD = ABE$$
$$CE = ABD$$
$$DE = ABC$$

The design in Table 5.6 is called a resolution V design, because the shortest word in the defining relation (other than I) has 5 letters. Therefore, every main effect is confounded with a four-factor interaction, and every two-factor interaction is confounded with a three-factor interaction. Therefore, if three and four factor interactions can be assumed to be negligible, a five factor experiment can be completed with only 16 runs, and all main effects and two-factor interactions will be estimable.

A half-fraction design and the alias pattern for that design can be created easily with the FrF2() function in the R package FrF2(). The example code below will create the design in Table 5.6 and its alias pattern.

```
R>library(FrF2)
R>design3<-FrF2(16,5,generators=c("ABCD"),
   randomize=F)
R>dmod<-lm(rnorm(16)~(A*B*C*D*E),data=design3)
R>aliases(dmod)
```

The first argument to the FrF2() function call is the number of runs in the design, and the second argument is the number of factors. The default factor names for the FrF2() are A, B, C, D, and E. The factor names and levels can be changed by including a factor.names argument as shown in the code that was used to analyze the data from the battery assembly experiment in the last section. The argument generators=c("ABCD") specifies that the fifth factor E will be assigned to the $ABCD$ interaction column. This argument is optional, and if left off, the FrF2() will always create the column for the fifth factor by setting it equal to the highest order interaction among the first four factors.

To get the alias pattern for the design we have to fit a model. Since the experiments have not yet been run and the design is being evaluated, a vector of 16 random normal observations were created, with the `rnorm` function in R, and used as the response data for the model `dmod`. The model, `(A*B*C*D*E)`, in the `lm()` function call above specifies that the saturated model is fit including all interactions up to the 5-way. The argument `data=design3`, and the function call `aliases(dmod)` will reproduce the alias pattern shown in Equation 5.7.

5.4.2   Example of a One-half Fraction of a 2^5 Designs

This example is patterned after the famous study made by Taguchi and Wu[95], and incorporating suggestions by Box Bisgaard and Fung[10]. In 1950 the Ina tile company in Japan was experiencing too much variability in the size of the tiles. A known common cause was the problem. There was a temperature gradient from the center to the walls of the tunnel kiln used for firing the tiles. This caused the dimension of the tiles produced near the walls to be somewhat smaller than those in the center of the kiln.

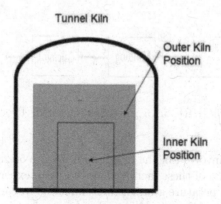

FIGURE 5.11: Tunnel Kiln

Unfortunately in 1950 the Ina tile company could not afford to install instrumentation to control the temperature gradient in their kiln. Therefore, Taguchi called the temperature gradient a noise factor that could not be controlled. There were other factors that could be easily controlled. Taguchi sought to find a combination of settings of these controllable factors, through use of an experimental design, that could mitigate the influence of the noise factor. This is referred to as a Robust Design Experiment. Figure 5.12

diagrammatically illustrates the factors and response for a Robust Design Experiment.

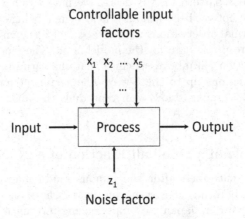

FIGURE 5.12: Robust Design

Figure 5.13 is a flow diagram of the process to make tiles at the Ina tile company

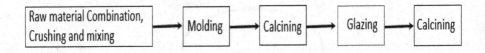

FIGURE 5.13: Tile Manufacturing Process

The variables involved in the first step in Figure 5.12 could be easily controlled. Therefore, 5 of these factors were chosen to experiment with, and the uncontrollable temperature gradient was simulated by sampling tiles from the center and outsides of the tunnel kiln, then measuring their dimensions. Table 5.7 shows the factors and levels for the Robust Design Experiment.

A 2^{5-1} fractional factorial experiment was used with the generator $E = ABCD$. Table 5.8 shows the list of experiments in standard order and the resulting average tile dimensions. Each experiment in the 2^{5-1} design was repeated at both levels of factor F. By doing this the design changes from a 2^{5-1} to a 2^{6-1}, and the defining relation changes from $I = ABCDE$ to $I = -ABCDEF$.

TABLE 5.7: Factors and Levels in Robust Design Experiment

Controllable Factors	Level $(-)$	Level $(+)$
A–kind of agalmatolite	existing	new
B–fineness of additive	courser	finer
C–content of waste return	0%	4%
D–content of lime	1%	5%
E–content of feldspar	0%	5%
Noise Factor	Level $(-)$	Level $(+)$
F–furnace position of tile	center	outside

TABLE 5.8: List of Experiments and Results for Robust Design Experiment

Random Run No.	A	B	C	D	E	F=Furnace Position center	outside
15	$-$	$-$	$-$	$-$	$+$	10.14672	10.14057
13	$+$	$-$	$-$	$-$	$-$	10.18401	10.15061
11	$-$	$+$	$-$	$-$	$-$	10.15383	10.15888
1	$+$	$+$	$-$	$-$	$+$	10.14803	10.13772
10	$-$	$-$	$+$	$-$	$-$	10.15425	10.15794
2	$+$	$-$	$+$	$-$	$+$	10.16879	10.15545
3	$-$	$+$	$+$	$-$	$+$	10.16728	10.15628
12	$+$	$+$	$+$	$-$	$-$	10.16039	10.17175
16	$-$	$-$	$-$	$+$	$-$	10.17273	10.12570
8	$+$	$-$	$-$	$+$	$+$	10.16888	10.13028
9	$-$	$+$	$-$	$+$	$+$	10.19741	10.15836
14	$+$	$+$	$-$	$+$	$-$	10.19518	10.14300
6	$-$	$-$	$+$	$+$	$+$	10.17892	10.13132
5	$+$	$-$	$+$	$+$	$-$	10.16295	10.12587
7	$-$	$+$	$+$	$+$	$-$	10.19351	10.13694
4	$+$	$+$	$+$	$+$	$+$	10.19278	10.11500

In the R code shown below, the FrF2() function was used to create the design in the standard order, and the vector y contains the data. Since the design is resolution VI, all main effects and two-factor interactions can be estimated if four-factor and higher order interactions are assumed to be negligible. There are only 6 main effects and $\binom{6}{2} = 15$ two-factor interactions but 32 observations. Therefore the command mod<-lm(y~(.)^3,data=DesF) was used to fit the saturated model which includes 10 pairs of confounded or aliased three-factor interactions in addition to the main effects and two-factor interactions. The aliases(mod) shows the alias for each three-factor interaction estimated, and the output of this command is shown below the code.

```
R>library(FrF2)
R>DesF<-FrF2(32,factor.names=list(A="",B="",C="",D="",
  F="",E=""),generators=list(c(1,2,3,4,5)),
  randomize=FALSE)
R>y<-c(10.14057,10.15061,10.15888,10.13772,10.15794,
       10.15545,10.15628,10.17175,10.14672,10.18401,
       10.15383,10.14803,10.15425,10.16879,10.16728,
       10.16039,10.12570,10.13028,10.15836,10.14300,
       10.13132,10.12587,10.13694,10.11500,10.17273,
       10.16888,10.19741,10.19518,10.17892,10.16295,
       10.19351,10.19278)
R>mod<-lm(y~(.)^3,data=DesF)
R>aliases(mod)


A:B:C = -D:F:E
A:B:D = -C:F:E
A:B:F = -C:D:E
A:B:E = -C:D:F
A:C:D = -B:F:E
A:C:F = -B:D:E
A:C:E = -B:D:F
A:D:F = -B:C:E
A:D:E = -B:C:F
A:F:E = -B:C:D


R>summary(mod)
R>library(daewr)
R>fullnormal(coef(mod)[2:32],alpha=.01,refline="FALSE")
R>interaction.plot(Furnace_position,Content_Lime,resp)
```

fullnormal(coef(mod)[2:32],alpha=.01,refline="FALSE") creates a full normal plot of the 31 estimable effects from the 32-run design. This plot is shown in Figure 5.14. It shows that the only significant effects appear to be main effect D-the content of lime and the interaction between factor D and the noise factor F-furnace position.

The interaction.plot(Furnace_position,Content_Lime,resp) command creates the interaction plot shown in Figure 5.15 (minus the enhancements that were added later). In this plot the noise factor is on the horizontal axis and the two lines represent the conditional main effect of F-furnace position for each level of D-content of lime. The plot shows that Furnace position effect is much stronger when there is only 1% of lime in the raw materials.

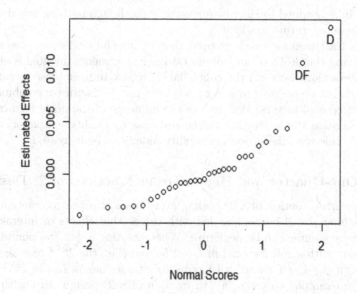

FIGURE 5.14: Normal Plot of Half-Effects

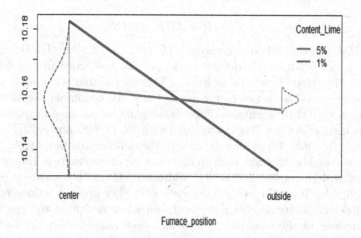

FIGURE 5.15: CD Interaction Plot

By adding 5% lime to the raw materials the effect of the furnace position (or temperature gradient) is greatly reduced. The red and green normal curves on the left and right of the lines representing the conditional effect of furnace position illustrate the variability in tile size expected to be caused by the uncontrollable temperature gradient. That expected variation is much less when 5% lime is added to the raw materials even though nothing was done to control the temperature gradient.

In this fractional factorial example, discovering the interaction was again key to solving the problem. In Robust Design Experiments the goal is always to find interactions between the controllable factors and the noise factors. In that way it may be possible to select a level of a control factor (or combination of levels of control factors) that reduce the influence of the noise factors and thereby decrease the variability in the response or quality characteristic of interest. In this way, the process capability index can be increased.

5.4.3   One-Quarter and Higher Order Fractions of 2^k Designs

In a one-quarter fraction of a 2^k design, every main effect or interaction that can be estimated will be confounded with three other effects or interactions that must be assumed to be negligible. When creating a $\frac{1}{4}$th fraction of a 2^k design, start with a full factorial in $k-2$ factors (i.e., the 2^{k-2} base design). Next, assign the $k-1$st and kth factors to interactions in in the 2^{k-2} base design. For example, to create a $\frac{1}{4}$th fraction of a 2^5 design start with a 2^3 8-run base design. Assign the 4th factor to the ABC interaction in the base design, and the 5th factor to one of the interactions AB, AC, or BC. If the 5th factor is assigned to AB, then the defining relation for the design will be:

$$I = ABCD = ABE = CDE, \tag{5.8}$$

where CDE is found when multiplying $ABCD \times ABE = A^2B^2CDE$

This is a resolution III design, since the shortest word in the defining relation (other than I) has three letters. This means that some main effects will be confounded with two-factor interactions. All estimable effects will be confounded with three additional effects that must be assumed negligible. For example, main effect A will be confounded with BCD, BE, and $ACDE$, which were found by multiplying all four sides of the defining relation by A.

In a one-eighth fraction each main effect or or interaction that can be estimated will be confounded with 7 additional effects that must be considered negligible. When creating one-fourth or higher order fractions of a 2^k factorial design, extreme care must be taken when assigning the generators or interactions in the base design to which additional factors are assigned. Otherwise, two main effects may be confounded with each other. When using the FrF2() function to create a higher order fractional design, leave off the argument generators=c( ). The FrF2() function has an algorithm that will produce the *minimum aberration-maximum resolution* design for the fraction size defined by the number of runs and number of factors. This design will have the least confounding among main effects and low order interactions that is possible.

5.4.4 An Example of a $\frac{1}{8}$th Fraction of a 2^7 Design

Large variation in the viscosity measurements of a chemical product were observed in an analytical laboratory Snee[88]. The viscosity was a key quality characteristic of a high volume product. Since it was impossible to control viscosity without being able to measure it accurately, it was decided to conduct a ruggedness test (Youden and Steiner[41], and Wernimount[99]) of the measurement process to see which variables if any influenced the viscosity measurement. Discussions about how the measurement process worked identified 7 possible factors that could be important. A description of the measurement process and possible factors follows.

"The sample was prepared by one of two methods (M1,M2) using moisture measurements made on a volume or weight basis. The sample was then put into a machine and mixed at a given speed (800–1600 rpm) for a specified time period (0.5-3 hrs.) and allowed to "heal" for one to two hours. The levels of the variables A-E shown in Table 5.9 were those routinely used by laboratory personnel making the viscosity measurements. There were also two Spindles (S1,S2) used in the mixer that were thought to be identical. The apparatus also had a protective lid for safety purposes and it was decided to run tests with and without the lid to see if it had any effect" (Snee[88]).

TABLE 5.9: Factors and Levels in Ruggedness Experiment

Factors	Level $(-)$	Level $(+)$
A–Sample Preparation	M1	M2
B–Moisture measurement	Volume	Weight
C–Mixing Speed(rpm)	800	1600
D–Mixing Time(hrs)	0.5	3
E–Healing Time(hrs)	1	2
F–Spindle	S1	S2
G–Protective Lid	Absent	Present

A one-eighth fraction (2^{7-3}) was constructed with generators $E = BCD$, $F = ACD$, and $G = ABC$. Samples of product from a common source were tested at the 16 conditions in the design. If the measurement process were rugged, then the differences in the measurements should have been due only to random measurement error.

To match the design in the article, the arguments generators=c("BCD", "ACD","ABC") and randomize=FALSE were used in the call to the FrF2 function call that created the design as shown in the R code and output below.

```
R>library(FrF2)
R>frac1<-FrF2(16,7,generators=c("BCD","ACD","ABC"),
        randomize=FALSE)
R>frac1
```

```
     A   B   C   D   E   F   G
1   -1  -1  -1  -1  -1  -1  -1
2    1  -1  -1  -1  -1   1   1
3   -1   1  -1  -1   1  -1   1
4    1   1  -1  -1   1   1  -1
5   -1  -1   1  -1   1   1   1
6    1  -1   1  -1   1  -1  -1
7   -1   1   1  -1  -1   1  -1
8    1   1   1  -1  -1  -1   1
9   -1  -1  -1   1   1   1  -1
10   1  -1  -1   1   1  -1   1
11  -1   1  -1   1  -1   1   1
12   1   1  -1   1  -1  -1  -1
13  -1  -1   1   1  -1  -1   1
14   1  -1   1   1  -1   1  -1
15  -1   1   1   1   1  -1  -1
16   1   1   1   1   1   1   1
class=design, type= FrF2.generators
```

The viscosity measurements obtained after running the experiments and the alias pattern for the design (including only up to 4 factor interactions) is shown in the code examples below along with the calculated effects.

```
R>library(FrF2)
R>library(daewr)
R>library(FrF2)
R>viscosity<-c(2796,2460,2904,2320,2800,3772,2420,3376,
              2220,2548,2080,2464,3216,2380,3196,2340)
R> aliases(lm(viscosity~(.)^4, data=frac1))

A = B:C:G = B:E:F = C:D:F = D:E:G
B = A:C:G = A:E:F = C:D:E = D:F:G
C = A:B:G = A:D:F = B:D:E = E:F:G
D = A:C:F = A:E:G = B:C:E = B:F:G
E = A:B:F = A:D:G = B:C:D = C:F:G
F = A:B:E = A:C:D = B:D:G = C:E:G
G = A:D:E = B:D:F = C:E:F = A:B:C
A:B = A:C:D:E = A:D:F:G = B:C:D:F = B:D:E:G = C:G = E:F
A:C = A:B:D:E = A:E:F:G = B:C:E:F = C:D:E:G = B:G = D:F
A:D = A:B:C:E = A:B:F:G = B:C:D:G = B:D:E:F = C:F = E:G
A:E = A:B:C:D = A:C:F:G = B:C:E:G = C:D:E:F = B:F = D:G
A:F = A:B:D:G = A:C:E:G = B:C:F:G = D:E:F:G = B:E = C:D
A:G = A:B:D:F = A:C:E:F = B:E:F:G = C:D:F:G = B:C = D:E
B:D = A:B:C:F = A:B:E:G = A:C:D:G = A:D:E:F = C:E = F:G
A:B:D = A:C:E = A:F:G = B:C:F = B:E:G = C:D:G = D:E:F
```

```
R>modf<-lm(viscosity~A+B+C+D+E+F+G+A:B+A:C+A:D+A:E+
      A:F+A:G+B:D+A:B:D,data=frac1)
R>summary(modf)
R>fullnormal(coef(modf)[-1],alpha=.15)
```

```
Coefficients:
            Estimate Std. Error t value Pr(>|t|)
(Intercept)  2705.75         NA      NA       NA
A1               1.75         NA      NA       NA
B1             -68.25         NA      NA       NA
C1             231.75         NA      NA       NA
D1            -150.25         NA      NA       NA
E1              56.75         NA      NA       NA
F1            -328.25         NA      NA       NA
G1               9.75         NA      NA       NA
A1:B1          -14.25         NA      NA       NA
A1:C1           27.75         NA      NA       NA
A1:D1         -124.25         NA      NA       NA
A1:E1          -19.25         NA      NA       NA
A1:F1           -4.25         NA      NA       NA
A1:G1          -36.25         NA      NA       NA
B1:D1           32.75         NA      NA       NA
A1:B1:D1        18.75         NA      NA       NA

Residual standard error: NaN on 0 degrees of freedom
Multiple R-squared:       1,Adjusted R-squared:      NaN
F-statistic:   NaN on 15 and 0 DF,  p-value: NA
```

There are 2^{7-3} effects and interactions that were estimated from this design, and each effect or interaction that can be estimated is confounded with 7 effects or interactions which must be assumed to be negligible. Due to this confounding, it would appear to be a difficult task to identify which factors and interactions actually had significant effects on the viscosity measurement. However, three principles that have been established by experience simplify the task. The first principle is the *effect sparsity principle*, the second is the *hierarchical ordering principle* (which was discussed earlier), and the third is the *effect-heredity principle*.

The effect sparsity principle implies that when many factors are studied in a factorial design, usually only a small fraction of them will be found to be significant. Therefore, the normal probability plot of effects, shown in Figure 5.16 is a good way to identify the subset of significant effects and interaction effects.

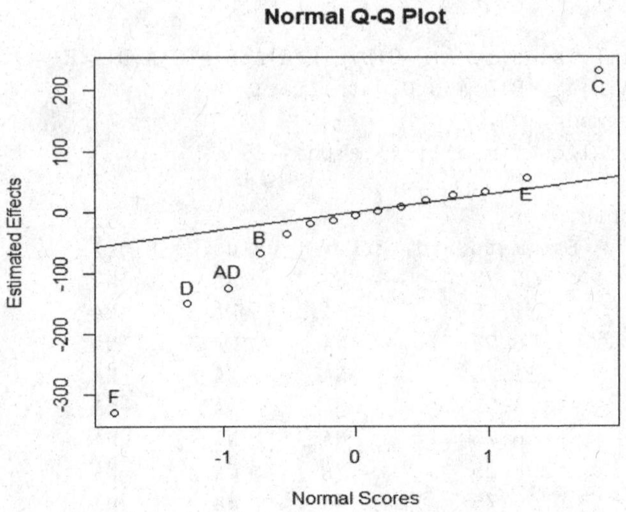

FIGURE 5.16: Normal Plot of Effects from Ruggedness Test

Recall that the hierarchical ordering principle (that was defined earlier) states that main effects and low order interactions (like two-factor interactions) are more likely to be significant than higher order interactions (like 4 or 5-way interactions etc.).

The effect-heredity principle states that interaction effects among factors that have insignificant main effects are rare, and that interactions involving factors that have significant main effects effects are much more likely to occur.

These three principles will be illustrated by determining which factors and interactions actually had significant effects on the viscosity measurement.

The normal plot of the half-effects in Figure 5.16 shows only 5 of the 15 calculated effects appear to be significant. They are the average main effects for C=Mixing speed, F=Spindle, D=Mixing Time, B=Moisture measurement, as well as the AD interaction (A=Sample Preparation method). Each of the main effects is confounded with four three-factor interactions, and three five-factor interactions as shown in the alias pattern given above. The AD interaction is confounded with four four-factor interactions and two other two-factor interactions.

The hierarchical ordering principle leads to the belief that the main effects (rather than three- or five-factor interactions) caused the significance of the C, F, D, B, and E half-effects, and that a two-factor interaction rather that one of the four-factor interactions confounded with it caused the significance of AD. However, the hierarchical ordering principle does not help in determining which of the two-factor interactions AD=CF=EG caused the significance of AD.

The two largest effects (in absolute value) are C=Mixing Speed, and F=Spindle, therefore the effect-heredity principle leads to the belief that the CF interaction is the cause of the confounded set of two-factor interactions AD=CF=EG.

The R code below produces the CF interaction plot shown in Figure 5.17, where it can be seen that C=Mixing speed, appears to have a larger effect on the viscosity measurement when Spindle 1 is used rather than when Spindle 2 is used.

```
R>library(FrF2)
R>Mixing_speed<-rep(0,16)
R>Spindle<-rep(1,16)
R>  for ( i in 1:16) {
R>  if(frac1$C[i]==-1) {Mixing_speed[i]=800}
R>  else  {Mixing_speed[i]=1600}
R>  if(frac1$F[i]==1) Spindle[i]=2
R>                        }
R>interaction.plot(Mixing_speed,Spindle,viscosity,
   type="b",pch=c(1,2),col=c(1,2))
```

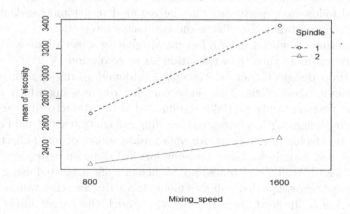

FIGURE 5.17: CF Interaction Plot

Additional experiments confirmed that the CF interaction, rather than AD or EG, was causing the large differences in measured viscosity as seen in Figure 5.17. It was clear from the 2^{7-3} experiments that the differences in measured viscosity were due to more-than-random measurement error. The two Spindles (factor F) that were thought to be identical were not identical, and on the average viscosity measurements made using Spindle 1 were $2\times$ 328.25=656.6 greater than viscosity measurements made using Spindle 2. In addition, the difference in viscosity measurements made using Spindle 1 and Spindle 2 was greater when the mixing speed was 1600 rather than 800.

The results of the experimentation showed the measurement process was not rugged, and that in the future the settings for the five factors B=Moisture measurement, C=Mixing speed, D=Mixing time, E=Healing time, and F=Spindle needed to be standardized and tightly controlled in order to reduce the measurement precision.

The examples in the last two sections of this chapter illustrate the fact that interactions are common in experiments. Therefore, full-factorial experiments like those shown in Section 5.3, or regular fractions of 2^k designs like those shown in this section (Section 5.4), are very useful in most trouble-shooting or process-improvement projects.

5.5   Alternative Screening Designs

In resolution III fractional factorial 2^{k-p} designs some main effects will be completely confounded with two-factor interactions, and in resolution IV 2^{k-p} designs some two-factor interactions will be completely confounded with other two-factor interactions. The confounding makes the analysis of these designs tricky, and follow-up experiments may be required to get independent estimates of the important main effects and two-factor interactions.

For experiments with 6, 7, or 8 factors, alternative screening designs in 16 runs require no more runs than resolution III or resolution IV 2^{k-p} designs. However, these designs do not have complete confounding among main effects and two-factor interactions. They have complex confounding which means that main effects are only partially confounded with two-factor interactions. Models fit to designs with complex confounding can contain some main effects and some two-factor interactions. An appropriate subset of main effects and interactions can usually be found be using forward stepwise regression.

Hamada and Wu[35] and Jones and Nachtsheim[48] suggested using a forward stepwise regression that enforces model hierarchy. In other words, when an interaction is the next term to enter the model, the parent linear main effect(s) are automatically included, if they are not already in the model. The number of steps in the forward selection procedure is the number of steps until the last term entered is still significant. When the term entered at the current step is not significant, revert to the previous step to obtain the final model. This procedure greatly simplifies the model fitting for alternative screening designs.

The Altscreen() function in the daewr package can be used to recall 16 run alternative screening designs from Jones and Montgomery[47]. The

`HierAFS()` function in the same package can be used to fit a model using a forward stepwise regression that enforces model hierarchy.

5.5.1   Example

Montgomery[71] presented an example of a 16-run 2_{IV}^{6-2} fractional factorial experiment for studying factors that affected the thickness of a photoresist layer applied to a silicon wafer in a spin-coater. The factors and levels are shown in Table 5.10

TABLE 5.10: Factors and Levels in Spin-Coater Experiment

Factors	Level $(-)$	Level $(+)$
A–spin speed	−	+
B–acceleration	−	+
C–volume of resist applied	−	+
D–spin time	−	+
E–resist viscosity	−	+
F–exhaust rate	−	+

In this example, the analysis showed that main effects A, B, C, and E, along with the confounded string of two-factor interactions AB=CE, were significant. However, since both main effects A and B and main effects C and E were significant, the effect-heredity principle did not help in trying to determine whether the CE interaction or the AB interaction was causing the significance of the confounded string. This situation is different from the last example. In that example, main effects C, F, and D were significant along with the confounded string of two-factor interaction AD=CF=EG. Since C and F were the largest main effects, the effect-heredity principle would lead one to believe that CF was causing the significance of the confounded string of interactions.

Since neither the hierarchical ordering principle nor the effect-heredity principle made it clear what interaction was causing the significance of the confounded string AB=CE in the present example, 16 additional follow-up experiments were run to determine which interaction was causing the significance and to identify the optimal factor settings. The follow-up experiments were selected according to a fold-over plan Lawson[57], and the analysis of the combined set of 32 experiments showed that the four main effects found in the analysis of the original 16 experiments were significant along with the CE interaction.

To demonstrate the value of an Alternative Screening Design, Johnson and Jones[45] considered the same situation as this example. They used the model found after analysis of the combined 32 experiments and simulated data for the photo-resist thickness. In the code below, the `Altscreen()` function in the **daewr** package was used to generate the same six-factor alternative screening design used in Johnson and Jones' article, and the thickness data was their simulated data. The `HierAFS()` function identified the correct model for the simulated data in three steps, and the variable that would have entered in the fourth step, BE, was not significant (p-value=0.11327). Had the original experiment been conducted with this design, no follow-up experiments would have been required, and the conclusions would have been reached with half the number of experiments.

```
R>library(daewr)
R>Design<-Altscreen(nfac=6,randomize=FALSE)
R>Thickness<-c(4494,4592,4357,4489,4513,4483,4288,
              4448,4691,4671,4219,4271,4530,4632,
              4337,4391)
R>cat("Table of Design and Response")
R>cbind(Design,Thickness)
R>HierAFS(Thickness,Design,m=0,c=6,step=4)
   formula               R2
y~A                    0.660
y~B+A                  0.791
y~C+E+C:E+B+A          0.953
y~B+E+B:E+C+C:E+A      0.965
```

The R code below shows the calculated effects (from the `lm()` function) for the model found by Johnson and Jones[45] and by using three steps of the `HierAFS()` function. The assumptions of the least-squares fit can be checked using the `plot.lm()` function, as shown in the block of code above Figure 5.7.

```
R> mod<-lm(Thickness~A+B+C+E+C:E,data=Design)
R> summary(mod)
Call:
lm(formula = Thickness ~ A + B + C + E + C:E, data = Design)

Residuals:
    Min      1Q  Median      3Q     Max
-56.625 -19.625   2.625  23.125  53.875

Coefficients:
             Estimate Std. Error t value Pr(>|t|)
(Intercept) 4462.875      9.494 470.073  < 2e-16 ***
A             85.500     11.628   7.353 2.44e-05 ***
B            -77.750     11.628  -6.687 5.45e-05 ***
C            -34.250      9.494  -3.608  0.00479 **
E             21.500      9.494   2.265  0.04700 *
C:E           54.750     13.427   4.078  0.00222 **
---
```

```
Signif. codes:  0 '***' 0.001 '**' 0.01 '*' 0.05 '.' 0.1 ' ' 1

Residual standard error: 37.98 on 10 degrees of freedom
Multiple R-squared:  0.9533,Adjusted R-squared:    0.93
F-statistic: 40.85 on 5 and 10 DF,  p-value: 2.447e-06
```

A=spin speed had a positive effect on thickness, and increasing spin speed increased thickness. B=acceleration had a negative effect on thickness, and increasing acceleration decreased thickness. Because of the significant interaction, it is difficult to say from the main effects what combinations of C=volume of resist applied, and E=resist viscosity would increase or decrease the thickness. The code below produces the interaction plot shown in Figure 5.18. This plot makes it clear.

```
R>resist_viscosity<-Design$E
R>interaction.plot(Design$C,resist_viscosity,Thickness,
   type="b",xlab="volume of resist applied",
   pch=c(1,2),col=c(1,2))
```

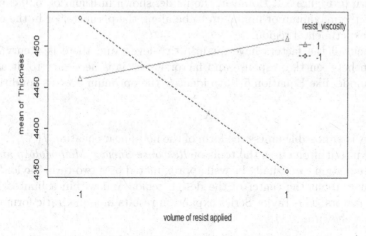

FIGURE 5.18: CE Interaction Plot

The greatest thickness results with a low C=volume of resist applied and a low E=resist viscosity, and the smallest thickness is produced using low E=resist viscosity but a high C=volume of resist applied.

5.6   Response Surface and Definitive Screening Experiments

The prediction model for a 2^k factorial design, a 2^{k-p} fractional factorial design, or an alternative screening design that only includes the significant main effects and two-factor interactions, would include a subset of the terms in Equation 5.9.

$$y = \beta_0 + \sum_{i=1}^{k} \beta_i X_i + \sum_{i<j}^{k} \sum^{k} \beta_{ij} X_i X_j + \epsilon, \tag{5.9}$$

where X_i for $i = 1 \ldots, k$ are the coded and scaled factor levels from Equation 5.2.

This equation will make accurate predictions of the response y between the high and low levels of all factors as long as the relationship between the factor levels and response y are approximately linear. If there is a curvilinear relation between the response and factor levels, a maximum, minimum, or a desired target level of the response may be interior to the experimental region, as shown in Figure 5.1. However, the model shown in Equation 5.9, can only predict the maximum or minimum to be along the perimeter or in the corner of the experimental region.

When all the factors have quantitative levels and there is a curvilinear relation between the response and factor levels, it is necessary to fit a non-linear model like Equation 5.10 to identify the optimum process conditions.

$$y = f(X_1, \ldots, X_k) + \epsilon. \tag{5.10}$$

This is impossible unless the form of the non-linear equation $f(X_1, \ldots, X_k)$ is known, but if one uses the tools of *Response Surface Methodology* any unknown equation can usually be well approximated by a two-term Taylor Series expansion about the center of the design region and within a limited range on the factors. The Taylor Series expansion results in a quadratic form of the equation shown in 5.11.

$$y = \beta_0 + \sum_{i=1}^{k} \beta_i X_i + \sum_{i=1}^{k} \beta_{ii} X_i^2 + \sum_{i<j}^{k} \sum^{k} \beta_{ij} X_i X_j + \epsilon, \tag{5.11}$$

where again X_i for $i = 1 \ldots, k$ are the coded and scaled factor levels from Equation 5.2. Figure 5.19 is a representation of one type quadratic equation in two factors X_1 and X_2. In this figure it can be seen that the minimum response lies within the experimental region near the mid-level of X_1 and toward the low level of X_2. The model for a factorial experiment like Equation 5.1 (represented by the plane above the surface in Figure 5.19) would be incapable of predicting the location of the minimum response. For this reason the ISO Technical

Committee 69 recognized the importance of response surface experimentation for process improvement studies, in addition to 2^k factorial and 2^{k-p} fractional factorial designs (Johnson and Boulanger[44]).

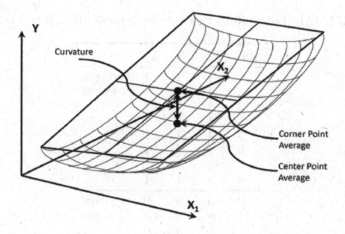

FIGURE 5.19: Quadratic Response Surface

In order to fit a quadratic equation to the data obtained from experiments, there must be at least three levels of each factor rather than the two levels contained in the 2^k factorial designs, 2^{k-p} fractional factorial designs, or alternative screening designs.

One classical experimental design used in Response Surface Methodology is the Central Composite Design. It has five levels for each factor and is composed of a 2^k design, plus a center point at the center-level of each factor, plus axial or star points that are outside the 2^k region. Figure 5.20 shows a visual representation of the experiments for a Central Composite Design in three factors.

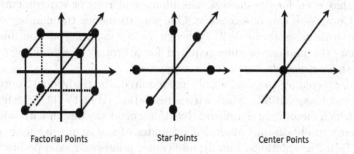

FIGURE 5.20: Central Composite Design

Table 5.11 is a list of the runs in a Central Composite Design in three coded and scaled factors X_i, for $i = 1, \ldots, 3$. In this table $\alpha = \sqrt{3}$, and is outside the coded low-high range of -1 to 1. In general $\alpha = \sqrt{k}$, where k is the number of factors in the design.

TABLE 5.11: List of Runs for Central Composite Design in 3 Factors

run	X_1	X_2	X_3	
1	-1	-1	-1	Factorial
2	1	-1	-1	Points
3	-1	1	-1	
4	1	1	-1	
5	-1	-1	1	
6	1	-1	1	
7	-1	1	1	
8	1	1	1	
9	$-\alpha$	0	0	axial
10	α	0	0	points
11	0	$-\alpha$	0	
12	0	α	0	
13	0	0	$-\alpha$	
14	0	0	α	
15	0	0	0	center
16	0	0	0	points
17	0	0	0	
18	0	0	0	
19	0	0	0	

Table 5.12 shows the total number of runs or experiments required for a Central Composite design that includes 5 center points. The first number 2^k is the number of factorial points, the second number $2 \times k$ is the number of axial points, and the third number is the number of center points. It can be seen that even for five factors, the number of runs or experiments for a Central Composite can be excessive. One way to reduce the number of runs is to substitute a resolution V 2^{k-p} design for the factorial portion, but even doing that, the number of runs required for a typical process improvement study with 7 or more factors can be impractical.

In order to reduce the number of runs required, the traditional approach is to experiment sequentially. Start with a resolution III or IV 2^{k-p} design, and identify which factors are significant. Next, augment the design with follow-up experiments to obtain unconfounded estimates of the significant main effects and two-factor interactions. Finally, add center points and axial points for the significant effects in order to fit a subset of the terms in Equation 5.11. Figure 5.21 outlines this strategy.

TABLE 5.12: Number of runs for Central Composite Design

Number of factors k	Number of runs for Central Composite Design
2	$2^2+4+5=13$
3	$2^3+6+5=19$
4	$2^4+8+5=29$
5	$2^5+10+5=47$
6	$2^6+12+5=81$
7	$2^7+14+5=157$

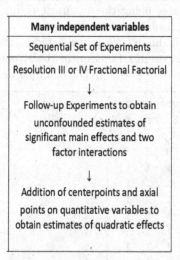

FIGURE 5.21: Sequential Experimentation Strategy

In 2011, Jones and Nachtsheim[48] proposed the use of Definitive Screening Designs that reduce the number of experiments required to fit a subset of Equation 5.11 in the significant factors. Their approach essentially combines an efficient screening design for identifying important factors and interactions, with center levels for the factors that allow fitting quadratic effects. These designs can be called super-saturated in that the number of runs in the design $(2 \times k + 1)$ is less than the number of coefficients $\left(1 + 2 \times k + \frac{k \times (k-1)}{2}\right)$ in Equation 5.11.

However, if there are several proposed factors in a process improvement study, it is unlikely that each factor will have significant linear, quadratic, and interaction terms with all the other factors. This is again because of the *effect sparsity principle*. When using a Definitive Screening Design, like an alternative screening design, an appropriate model consisting of the important subset of terms in Equation 5.11 can usually be obtained using a forward selection procedure.

When using a Definitive Screening Design, the selection procedure should take into account the special structure of the design (Jones and Nachtsheim[50]). The special structure of Definitive Screening Designs is that all linear main effects are completely unconfounded with all quadratic and two-factor interaction terms. Quadratic and two-factor interaction terms are partially confounded like they are in alternative screening designs. Therefore, when using a Definitive Screening Design, the important main effects can be determined first, and then, assuming strong effect heredity, a forward search is made with candidates being the quadratic terms and two-factor interactions involving the important main effects.

Jones and Nachtsheim's[48] paper listed a catalog of Definitive Screening Designs. The `DefScreen()` function in the R package `daewr` can be used to recall a design from this catalog. The example below illustrates this.

```
R>library(daewr)
R>Design<-DefScreen(m=8,c=0)
R>Design
      A  B  C  D  E  F  G  H
 1    0 -1  1  1 -1  1  1  1
 2    0  1 -1 -1  1 -1 -1 -1
 3   -1  0 -1  1  1  1  1 -1
 4    1  0  1 -1 -1 -1 -1  1
 5   -1 -1  0  1  1 -1 -1  1
 6    1  1  0 -1 -1  1  1 -1
 7    1 -1  1  0  1  1 -1 -1
 8   -1  1 -1  0 -1 -1  1  1
 9   -1 -1  1 -1  0 -1  1 -1
10    1  1 -1  1  0  1 -1  1
11    1 -1 -1 -1  1  0  1  1
12   -1  1  1  1 -1  0 -1 -1
13   -1  1  1 -1  1  1  0  1
14    1 -1 -1  1 -1 -1  0 -1
15    1  1  1  1  1 -1  1  0
16   -1 -1 -1 -1 -1  1 -1  0
17    0  0  0  0  0  0  0  0
```

The argument m=8 specifies that there are 8 quantitative 3-level factors in the design. The number of runs is $2 \times k + 1 = 17$, when $k = 8$. This design is listed in standard order since the default option `randomize=FALSE` was not overridden in the function call. When planning a Definitive Screening Design, the option `randomize=TRUE` should be used to recall the runs in a random order, and thus prevent biases from changes in unknown factors.

To illustrate the analysis of data from an Definitive Screening Design, consider the following example. Libbrecht et. al. [63] completed an 11-run Definitive Screening Design for optimizing the process of synthesizing soft template mesoporus carbons. Mesoporus carbons can be used for various applications such as electrode materials, absorbents, gas storage hosts, and support material for catalysts. The effect of five synthesis factors on the material properties of the carbons were considered. Because the relationship between the material properties, or responses, and the synthesis parameters, or factors, was expected to be nonlinear, it was desirable to use an experimental design that would allow fitting a subset of the terms in Equation 5.11. The five factors studied were: A (Carbonation temperature [°C]), B (Ratio of precursor/surfactant), C (EISA-Time [h]), D (Curing time [h]), and E (EISA surface area [cm^2]). Table 5.13 shows the range of each of the five factors.

TABLE 5.13: Factors and Uncoded Low and High Levels in Mesoporus Carbon Experiment

Factors	Level (−)	Level (+)
A–Carbonation temperature	600(°C)	1000(°C)
B–Ratio of precursor to surfactant	0.7	1.3
C–EISA-Time	6(h)	24(h)
D–Curing time	12(h)	24(h)
E–EISA surface area	65(cm^2)	306(cm^2)

The experimental outputs or properties of the mesoporus carbon were microporus and mesoporus surface area, pore volumes, and the weight % carbon. Among the purposes of the experimentation was to understand the effects of the acid catalyzed EISA synthesis parameters on the material properties and to find the optimal set of synthesis parameters to maximize mesopore surface area.

A traditional central composite design for five factors would entail 47 experiments. Since it was doubtful that all terms in a Equation 5.11 for $k =$ five factors would be significant, a Definitive Screening Design using less than one fourth that number of experiments (11) could be used.

The R code below shows the command to create the Definitive Screening Design using the `DefScreen()` function. The `randomize=FALSE` option was used for this experiment to create the design in standard order. The actual experiments were conducted in random order, but by creating the design in standard order, it was easier to pair the responses shown in the article to the experimental runs. The factor levels are in coded and scaled units. All Definitive Screening Designs in Jones and Nachtsheim's[48] catalog as recalled by the `DefScreen` function contain one center point where all factors are set at their mid levels. In this experiment, there were three center points, including additional center points allows for a lack of fit test to be performed. The argument `center=2` tells the function to add two additional center points.

```
R>library(daewr)
R>design<-DefScreen(m=5,c=0,center=2,
  randomize=FALSE)
R>design
      A  B  C  D  E
1     0  1  1 -1 -1
2     0 -1 -1  1  1
3     1  0 -1 -1  1
4    -1  0  1  1 -1
5     1 -1  0  1 -1
6    -1  1  0 -1  1
7     1 -1  1  0  1
8    -1  1 -1  0 -1
9     1  1  1  1  0
10   -1 -1 -1 -1  0
11    0  0  0  0  0
12    0  0  0  0  0
13    0  0  0  0  0
```

In the design listing shown above, the three center points (runs 11, 12, and 13) can be seen to have the center level (in coded and scaled units) for each factor.

The code below shows the responses recorded from the synthesis experiments, and the call of the function FitDefSc() used to fit a model to the data. The response, Smeso, is the mesoporus surface area. The option alpha=.05 controls the number of second-order terms that will be allowed in the model. A smaller the value of alpha is more stringent and allows fewer second-order terms to enter. A larger value of alpha is less stringent and allows more second-order terms to enter. Using the FitDefSc() values of alpha=.05 within the range .05 - .20 should be used.

```
R>library(daewr)
R>Smeso<-c(241,295,260,338,320,265,275,248,92.5,
          383,313,305,304)
R>FitDefSc(Smeso,design,alpha=.05)

Call:
lm(formula = y ~ (.), data = ndesign)

Residuals:
    Min      1Q  Median      3Q     Max
-13.667  -3.653   0.000   3.550  12.373
```

```
Coefficients:
             Estimate Std. Error t value Pr(>|t|)
(Intercept)  304.000       5.012  60.656 1.35e-09 ***
B            -76.030       4.362 -17.429 2.29e-06 ***
D            -26.609       4.533  -5.870 0.001082 **
B:D          -21.023       4.779  -4.399 0.004572 **
I(B^2)       -44.318       6.500  -6.819 0.000488 ***
A            -42.803       4.362  -9.812 6.45e-05 ***
E            -21.459       4.533  -4.734 0.003213 **
---
Signif. codes:  0 '***' 0.001 '**' 0.01 '*' 0.05 '.' 0.1 ' ' 1

Residual standard error: 11.21 on 6 degrees of freedom
Multiple R-squared:  0.9866, Adjusted R-squared:  0.9733
F-statistic:  73.9 on 6 and 6 DF,  p-value: 2.333e-05

             Sums of Squares  df F-value P-value
Lack of Fit        704.89318   4 7.24205 0.04059
Pure Error          48.66667   2
```

The lack of fit F-test shown at the bottom of the output is only produced when there is more than one center point in the design. The F-ratio compares the variability among the replicated center point responses, and the lack of fit variability. The lack of fit variability is the variability among the residuals (i.e., actual responses—fitted values) minus the variability among replicate center points. In this example, the F-test is significant. This means the difference in predictions from the fitted model and new experiments at the prediction conditions will be larger than replicate experiments at those conditions.

To improve the model, `alpha` was increased as shown in the code below to expand the model.

```
R>library(daewr)
R>FitDefSc(Smeso,design,alpha=.1)
Call:
lm(formula = y ~ (.), data = ndesign)

Residuals:
        1        2        3        4        5        6        7
 -0.39670 -2.32094  0.46480  3.15873 -0.12433  1.03021  3.54358
        8        9       10       11       12       13
  0.07995 -0.71346 -1.09831  4.45882 -3.54118 -4.54118
```

```
Coefficients:
             Estimate Std. Error t value Pr(>|t|)
(Intercept)   308.541       2.386 129.329 2.14e-08 ***
I(D^2)        -11.353       3.046  -3.727 0.020353 *
D             -26.609       1.855 -14.343 0.000137 ***
B             -76.030       1.785 -42.589 1.82e-06 ***
E             -21.459       1.855 -11.567 0.000319 ***
B:E            11.344       2.125   5.338 0.005933 **
B:D           -21.024       1.974 -10.649 0.000440 ***
I(B^2)        -37.509       2.974 -12.611 0.000228 ***
A             -42.803       1.785 -23.976 1.79e-05 ***
---
Signif. codes: 0 '***' 0.001 '**' 0.01 '*' 0.05 '.' 0.1 ' ' 1

Residual standard error: 4.586 on 4 degrees of freedom
Multiple R-squared:  0.9985,Adjusted R-squared:  0.9955
F-statistic: 334.9 on 8 and 4 DF,  p-value: 2.216e-05

          Sums of Squares  df F-value P-value
Lack of Fit       35.46992   2 0.72883 0.57842
Pure Error        48.66667   2
```

In the output it can be seen that two additional terms ((I(D^2), and B:E)) were added to the model and the lack of fit test is no longer significant. This means that predictions from this fitted model will be just as accurate as running additional experiments at the factor levels where the predictions are made.

Sometimes the FitDefSc() may include some terms in the model that are not significant at the $\alpha = .05$ level of significance. If that is the case, the lm() function can be used to refit the model eliminating the insignificant term with the largest p-value, as shown in the example in Section 5.5.1. This can be repeated until all remaining terms in the model are significant. After using the lm() function to fit a model, the assumptions of the least-squares fit can be checked by looking at the diagnostic plots produced by the plot.lm() functions as shown in the block of R code before Figure 5.7

Predictions from the fitted model can be explored graphically to identify the optimal set of synthesis parameters to maximize mesopore surface area, Smeso. Figure 5.22 shows contour plots of the predicted mesoporus surface area as a function of curing time and ratio of precursor/surfactant. The plot on the left is made when EISA surface area is at its low level of 65, and the plot on right is when EISA surface area is at its high level of 306. The code for producing these plots is in the R code for Chapter 5.

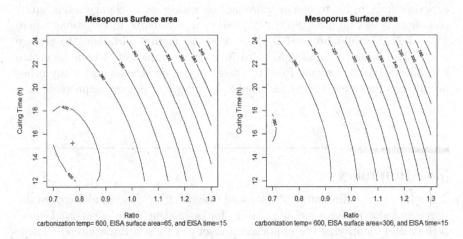

FIGURE 5.22: Contour Plots of Fitted Surface

It can be seen that the maximum predicted mesopore surface area is just over $400(\mathrm{m}^2/\mathrm{g})$ when carbonation temperature $= 600$, Ratio $\approx .80$, EISA time $= 15$, curing time ≈ 15, and EISA surface area $= 65$. These conclusions were reached after conducting only 13 experiments, far less than would be required using the traditional approach illustrated in Figure 5.21. The R function `ConstrOptim()` can also be used to get the optimal combination synthesis parameters numerically. For an example see Lawson[57].

5.7 Additional Reading

Plackett-Burman designs[79] and Model-Robust designs[62] are additional designs that have a minimal number of runs and complex confounding among two-level factor main effects and two-factor interactions. They are similar to alternative screening designs. They are also useful for identifying the important subset of terms in Equation 5.9 with fewer runs than traditional 2^k factorial designs. Jones and Nachtsheim[49] also developed a catalog of Definitive Screening Designs that contain three-level quantitative factors and up to 4 additional two-level categorical factors.

Blocked designs should be used to get more precise tests of main effects and interactions when there are uncontrollable conditions that cause extra variability in experimental units. Blocked designs group similar experimental units into more homogeneous blocks to reduce the variance within the blocks. When it is very difficult or expensive to completely randomize the order of experimentation due to one or more factors with hard to change levels, split-plot designs can be used. Split plot designs are available in situations where 2^k and 2^{k-p} would be used if there were no factors with hard to change levels. The book by Lawson[57] also includes examples of R code to create and analyze data resulting from the designs just mentioned and many other useful Experimental Designs for troubleshooting and process improvement.

5.8 Summary

This chapter has illustrated 2^k factorial designs, 2^{k-p} fractional factorial designs, and alternative screening designs for estimating main effects and interactions and identifying the important subset of terms in Equation 5.9. These designs are useful for identifying optimal processing conditions on a boundary or corner of the experimental region. Definitive Screening Designs were introduced as an efficient way of finding the important subset of terms in Equation 5.11, and for identifying optimal process conditions within the experimental region. There are many additional and useful experimental designs for estimating main effects, interaction effects, and non-linear effects with multiple factors.

5.9 Exercises

1. Run all the code examples in the chapter as you read.

2. Johnson and McNeilly[43] describe the use of a factorial experiment to quantify the impact of process inputs on outputs so that they can meet the customer specifications. The experiments were conducted at Metalor Technologies to understand its silver powder production process. High-purity silver powder is used in conductive pastes by the microelectronics industry. Physical properties of the powder (density and surface area in particular) are critical to its performance. The experiments set out to discover how three key input variables: (1) reaction temperature, (2) %ammonium, and (3) stir rate effect the density and surface area. The results of their experiments (in standard order) are shown in the following Tables. Two replicates were run at each of the eight treatment combinations in random order.

Factors	Level $(-)$	Level $(+)$
A=Ammonium	2%	30%
B=Stir Rate	100 rpm	150 rpm
C=Temperature	8°C	40°C

X_A	X_B	X_C	Density		Surface Area	
−	−	−	14.68	15.18	0.40	0.43
+	−	−	15.12	17.48	0.42	0.41
−	+	−	7.54	6.66	0.69	0.67
+	+	−	12.46	12.62	0.52	0.36
−	−	+	10.95	17.68	0.58	0.43
+	−	+	12.65	15.96	0.57	0.54
−	+	+	8.03	8.84	0.68	0.75
+	+	+	14.96	14.96	0.41	0.41

(a) Analyze the density response. Draw conclusions about the effects of the factors. Are there any significant interactions?

(b) Is the fitted model adequate?

(c) Make any plots that help you explain and interpret the results.

(d) Analyze the surface area response. Draw conclusions about the effects of the factors. Are there any significant interactions?

(e) Is the fitted model adequate?

(f) Make any plots that help you explain and interpret the results.

    (g) If the specifications require surface area to be between 0.3 and 0.6 cm2/g and density, must be less than 14 g/cm3, find operating conditions in terms of (1) reaction temperature, (2)

3. Create the minimum aberration maximum resolution (2^{5-2}) design.

4. Create a (2^{5-2}) design using the generators D=ABC, E=AC.

    (a) What is the defining relation for this design?

    (b) what is the alias pattern for this design?

    (c) Not knowing what factors will be represented by A-D, can you say whether the design you created in this exercise is any better or worse than the minimum aberration maximum resolution design you created in exercise 2?

5. Hill and Demier[39] showed an example of the use of the (2^{5-2}) design you created in exercise 3 for improving the yield of dyestuffs to make the product economically competitive. The product formation was thought to take place by the reaction steps shown.

$$A + B \longrightarrow C + \text{Others} \qquad\qquad (1)$$

$$C + F \xrightarrow{\text{B,E}} \text{Intermediates} + \text{Others} \qquad (2)$$

$$\text{Intermediates} \xrightarrow{\text{B,E}} \text{Product} + \text{Others} \qquad (3)$$

Where F is the basic starting material that reacts in the presence of a solvent with material C to give intermediates and then the final product. The second reaction is a condensation step, and the third reaction is a ring-closing step. It wasn't certain what the role of materials B and E played in reactions 2 and 3. Even though there were at least these three separate reactions, preliminary laboratory runs indicated that the reactants could be charged all at once. Therefore, the reaction procedure was to run for t1 hr at T1°C, corresponding to the steps needed for condensation, and then the temperature raised to T2°C for t2 more hours to complete the ring-closure. Preliminary experiments indicated that the time and temperature of the condensation deserved additional study, because this step appeared more critical and less robust than the ring-closing step. Other variables thought to affect the reaction were the amount of B, E, and the solvent. Therefore, these five variables were studied in an initial screening design to determine which had significant effects on product yield. The factors and levels were:

Factors	Level (−)	Level (+)
A=Condensation temp	90 °C	110 °C
B=Amount of B	29.3 cc	39.1 cc
C=Solvent volume	125 cc	175 cc
D=Condensation time	16 h	24 h
E=Amount of E	29 cc	43 cc

(a) Given that the observed yields and filtration times for the (2^{5-2}) experiment (in standard order) were:

Yields = 23.2,16.9,16.8,15.5,23.8,23.4,16.2,18

Filtration Time = 32,20,25,21,30,8,17,28

determine what factors and or interactions had significant effects on these responses, and interpret or explain them.

(b) The researchers believed that factor D=Condensation Temperature would have a large effect on both Yield and Filtration Time. The fact that it did not in these experiments, led them to believe that it may have a quadratic effect that was missed in the two-level experiments. What additional experiments would you recommend to check the possibility of a quadratic effect?

6. Compare the following two designs for studying the main effects and two-factor interactions among 6 factors. A=the minimum aberration maximum resolution 2^{6-1} and B=an Alternative Screening design for 6 factors. Create a two-by-two table with column headings for design A. and design B., and row headings Advantages and Disadvantages. Make entries in the four cells of the table showing the advantages and disadvantages for using each of the two designs for stated the purpose.

7. Compare the following two designs for studing the main effects and two-factor interactions among 7 factors. A=the minimum aberration maximum resolution 2^{7-3} and B=an Alternative Screening design for 7 factors. Create a two-by-two table with column headings for design A and design B, and row headings Advantages and Disadvantages. Make entries in the four cells of the table showing the advantages and disadvantages for using each of the two designs for stated the purpose.

(a) Which of the two designs would you recommend for the comparisom in question 6?

(b) Which of the two designs would you recommend for the comparisom in question 7?

8. Create a Definitive Screening Design for 6 three-level main effects that includes 3 additional center points.

(a) Create the design in standardized and randomized order.

(b) Assuming experiments had been conducted using the design you created in (a) and the responses in standard order were:

10.95,7.83,11.26,11.31,9.81,19.94,16.33,4.90,17.36,14.57,27.30,5.65,

9.27,9.30,8.77,11.11,

use the `FitDefSc()` function to find a model for the data.

(c) Does the model fit the data well? (i.e., is there any significant lack of fit?)

(d) If there are insignificant terms in the model you found, refit the model using the `lm()` eliminating one term at a time until all remaining terms are significant.

(e) Make the diagnostic plots to check the least-squares model assumptions for your final model.

6

Time Weighted Control Charts in Phase II

Shewhart control charts like the $\overline{X} - R$ charts discussed in Chapter 4 are very useful in Phase I. They can quickly detect a large shift in the mean or variance, and the patterns on the chart defined by the Western Electric rules mentioned in Section 4.2.3, are helpful in hypothesizing the cause of out-of-control conditions. If it can be demonstrated that a process is in "in-control" with a capability index of 1.5 or higher in Phase I, then there will be no more than 1,350 ppm (or 0.135%) nonconforming output in future production (as long as the process mean does not shift by more that 1.5 standard deviations). Therefore, to maintain a high quality level it is important that small shifts in the mean on the order of one standard deviation are quickly detected and corrected in during Phase II monitoring.

However, Shewhart charts are not sensitive to small shifts in the mean when used for Phase II process monitoring. When the only out-of-control signal used on a Shewhart chart is a point beyond the control limits, the OC may be too high and the ARL too long to quickly detect a small shift in the process mean. If additional sensitizing rules (like the Western Electric rules) are used in Phase II to decrease the out-of-control ARL, they also dramatically decrease the in-control ARL. This could cause unnecessary process adjustments or searches for causes. The solution is to use time weighted control charts. These charts plot the cumulative or weighted sums of all past observations, and are more likely to detect a small shift in the process mean.

6.1 Time Weighted Control Charts When the In-control μ and σ are known

To introduce the time weighted control charts, consider the random data produced by the R code below. The first 7 observations in the vector x1 are random observations from a normal distribution with mean $\mu = 50$ and standard deviation $\sigma = 5$. The next 8 observations in the vector x2 are random observations from the normal distribution with mean $\mu = 56.5$ and standard deviation $\sigma = 5$. Next, the code creates a Shewhart individuals chart of the data in both vectors assuming the known in-control mean and standard deviation are $\mu = 50$ and $\sigma = 5$. This chart is shown in Figure 6.1.

The vertical dashed line in the figure shows the point where the mean increased from 50 to 56.5. However, there are no points outside the control limits. It can be observed that all the points lie above the centerline on the right side of the chart.

```
R># random data from normal N(50,5) followed by N(56.6, 5)
R>set.seed(109)
R>x1<-rnorm(7,50,5)
R>set.seed(115)
R>x2<-rnorm(8,56.6,5)
R># individuals chart assuming mu=50, sigma=5
R>library(qcc)
R>qcc(x1,type="xbar.one",center=50,std.dev=5,
   newdata=x2)
```

This might suggest that the mean has increased, but, even using the Western Electric sensitizing rules, no out-of-control signal should be assumed until the 7th consecutive point above the center line occurs at the 14th point on the chart.

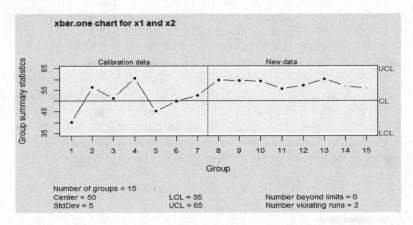

FIGURE 6.1: Individuals Chart of Random Data

To enhance the sensitivity of the control chart, without using the Western Electric rules that decrease the ARL_0 and increase the chance of a false positive, consider the cumulative sums of deviations from the centerline defined in Equation 6.1

$$C_i = \sum_{k=1}^{i}(X_k - \mu), \tag{6.1}$$

and shown in Table 6.1.

TABLE 6.1: Cumulative Sums of Deviations from the Mean

observation	value	Deviation from $\mu=50$	C_i
1	40.208	−9.792	−9.792
2	56.211	6.211	−3.581
3	51.236	1.236	−2.345
4	60.686	10.686	8.341
5	45.230	−4.770	3.571
6	49.849	−0.151	3.420
7	52.491	2.491	5.912
8	59.762	9.762	15.674
9	59.462	9.462	25.135
10	59.302	9.302	34.437
11	55.679	5.679	40.117
12	57.155	7.155	47.272
13	60.219	10.219	57.491
14	56.770	6.771	64.261
15	55.949	5.948	70.210

Figure 6.2 is a plot of the cumulative sums C_i by observation number i. In this figure, it can be seen that the cumulative sums of deviations from $\mu = 50$ remain close to zero for subgroups 1–7. These were generated from the normal distribution with mean 50. However, beginning at subgroup 8, where the process mean increased, the cumulative sums of deviations begin to increase noticeably.

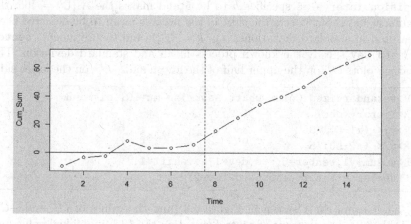

FIGURE 6.2: Plot of Cumulative Sums of Deviations from Table 6.1

This illustrates a time weighted weighted chart. Although the figure appears to show an out-of-control signal, there are no formal control limits on the graph. Using the tabular *Cusum control chart* is one way to obtain them.

6.1.1   Cusum Charts

Standardized tabular Cusum control charts work with individual measurements rather than subgroups. Two sets of cumulative sums of the standardized individual deviations from the mean are plotted on the tabular Cusum control chart. The first are the sums of the positive deviations, C_i^+ from the process mean, and the second set are the sums of the negative deviations C_i^-, from the process mean. The definition of these two cumulative sums are shown in Equation 6.2, and they are illustrated using the random data from Figure 6.1 and Table 6.2.

$$C_i^+ = max[0, y_i - k + C_{i-1}^+] \tag{6.2}$$

$$C_i^- = max[0, -k - y_i + C_{i-1}^-],$$

where $y_i = (x_i - \mu_0)/\sigma_0$, and $\mu_0 = 50$ and $\sigma_0 = 5$ are the known in-control process mean and standard deviation. The constant k is generally chosen to be $1/2$ the size of the mean shift in standard deviations that you would like to detect quickly. In order to detect a one standard deviation shift in the mean quickly, $k = 1/2$, and decision limits are placed on the chart at $\pm h$. An out-of-control signal occurs if the upper Cusum C_i^+ exceeds $+h$ or the lower Cusum C_i^+ falls below $-h$.

Using the cusum function in the qcc package, a Cusum chart of the random data is illustrated in the R code below and shown in Figure 6.3. The argument decision.interval=5 specifies h to be 5 and makes the $ARL_0 = 465$; the argument se.shift=1 sets the shift in the process mean to be detected as 1, measured in standard deviations, (i.e. $k = 1/2$); and the arguments center and std.dev specify the known process mean and standard deviation. The function plots C_i^+ on the upper half of the graph and $-C_i^-$ on the lower side.

```
R>#standardized Cusum chart assuming mu=50, sigma=5
R>library(qcc)
R>y1<-(x1-50)/5
R>y2<-(x2-50)/5
R>cusum(y1,center=0,std.dev=1,se.shift=1,
   decision.interval=5,newdata=y2)
```

The Cusum chart shown in Figure 6.3 (and the values of C_i^+ and $-C_i^-$ in Table 6.2) shows an out-of-control signal on the 12th individual value and indicates the process mean has increased.

If the Phase I OCAP indicates that the out-of-control signal can be corrected by a change in the level of a manipulable variable, then automated manufacturing processes often use the tabular Cusums in Equation 6.2 in a feedback loop. In this loop, the manipulable variable is automatically adjusted whenever an out-of-control signal is detected.

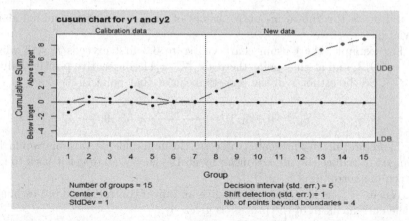

FIGURE 6.3: Standardized Cusum Chart of the Random Data in Table 6.1 with $k = 1/2$ and $h = 5$

TABLE 6.2: Standardized Tabular Cusums

Individual value	x_i	y_i	C_i^+	C_i^-
1	40.208	−1.958	0.000	−1.458
2	56.211	1.242	0.742	0.000
3	51.236	0.247	0.489	0.000
4	60.686	2.137	2.126	0.000
5	45.230	−0.954	0.673	−0.454
6	49.849	−0.030	0.142	0.000
7	52.491	0.498	0.141	0.000
8	59.762	1.952	1.593	0.000
9	59.462	1.892	2.985	0.000
10	59.302	1.860	4.346	0.000
11	55.679	1.136	4.982	0.000
12	57.155	1.431	5.913	0.000
13	60.219	2.043	7.457	0.000
14	56.770	1.354	8.311	0.000
15	55.949	1.190	9.000	0.000

When C_i^+ exceeds h, to give an out-of-control signal, an estimate of the current value of the process mean is given by Equation 6.3 .

$$\hat{\mu}_c = \mu + \sigma k + \frac{\sigma C_i^+}{N^+},\tag{6.3}$$

where $\hat{\mu}_c$ is the estimate of the current mean, N^+ is the consecutive number of nonzero values of C_i^+ up to and including the point where it exceeded h, and

μ, and σ are the known in-control values of the process mean and standard deviation.

For example, the Cusum chart in Figure 6.3 first exceeds $h = 5$ when $C_i^+ = 5.913$, and at that point there are $N^+ = 11$ consecutive positive values for C_i^+. So the estimate of the process mean at that point is:

$$\hat{\mu}_c = 50 + 5(.5) + \frac{(5)(5.913)}{11} = 55.19.$$

Therefore, in an automated process, the manipulable variable would be changed to reduce the process mean by $55.19 - 50 = 5.19$ to get it back to the known in-control level.

When $-C_i^-$ is less than $-h$, to give an out-of-control signal, an estimate of the current mean of the process is given by:

$$\hat{\mu}_c = \mu - \sigma k - \frac{\sigma C_i^-}{N^-}, \tag{6.4}$$

and the manipulable would be changed to increase the process mean back to the known in-control level.

6.1.1.1 Headstart Feature

If an out-of-control signal is detected in Phase II monitoring with a process that is not automated, or does not contain a manipulable variable with a known relationship to the process mean, then the process is usually stopped until a change or adjustment can be made. In this situation, it might be desirable to use a more sensitive control chart immediately after restarting the process, just in case the adjustment was not effective.

Lucas and Crosier[67] suggested a simple adjustment to the Cusum chart called the fast initial response (FIR) or headstart feature to make it more sensitive. Instead of starting the Cusum with $C_0^+ = 0.0$ and $C_0^- = 0.0$ as illustrated in Table 6.2, they suggested setting $C_0^+ = h/2$, and $C_0^- = -h/2$. This is illustrated with in Table 6.3 using the first four data points in Table 6.2, as if the process were just restarted. In this case $h = 5$, so $C_0^+ = 2.5$ and $C_0^- = -2.5$.

TABLE 6.3: Standardized Taular Cusums with FIR of $h/2$

Individual value	x_i	y_i	C_i^+	C_i^-
1	40.208	−1.958	0.042	-3.958
2	56.211	1.242	0.784	-2.216
3	51.236	0.247	0.531	-1.469
4	60.686	2.137	2.168	0.000

By using a headstart of $h/2$ and $h = 5$ on a Cusum chart, Lucas and

Crosier Lucas and Crosier[67] showed that the average run length (ARL) for detecting a one standard deviation change in the process mean was reduced from 10.4 to 6.35 (a 39% decrease). The ARL_0, or average run length for detecting a false out-of-control signal when the mean has not changed from the known in-control value, is only decreased by 7.5% (from 465 to 430).

6.1.1.2 ARL of Shewhart and Cusum Control Charts for Phase II Monitoring

If individual measurements are monitored rather than subgroups of data in Phase II monitoring, and a Shewhart chart is used, individual measurements would be plotted with the center line μ and the control limits $\mu \pm 3\sigma$, where μ and σ are the known values of the in-control process mean and standard deviation.

The OC or operating characteristic for the Shewhart chart is the probability of falling within the control limits, and it is a function of how far the mean has shifted from the known in-control mean (as discussed in Section 4.5.1). The average run length (ARL) is the expected value of a Geometric random with ARL= $1/(1 - OC)$. The OC for a Shewhart when the mean has shifted k standard deviations to the left or right can be calculated with the pnorm function in R as pnorm(3-k)-pnorm(-3-k)

Vance[97] published a computer program for finding the ARLs for Cusum control charts. The xcusum.arl function in the R package spc calculates ARLs for Cusum charts with or without the headstart feature by numerically approximating the integral equation used to find the expected or average run lengths. An example call of the xcusum.arl function is xcusum.arl(k=.5, h=5, mu = 1.0, hs=2.5, sided="two"). k=.5 specifies $k = 1/2$. h=5 specifies the decision interval $h = 5$. mu = 1.0 indicates that the ARL for detecting a one σ shift in the mean should be calculated. hs=2.5 specifies the headstart feature to be half the decision interval, and sided="two" indicates that the ARL for a two sided Cusum chart should be calculated.

The R code below was used to calculate the ARL for the Shewhart individuals chart for each of the mean shift values in the vector mu using the pnorm function; the ARL for the Cusum chart; and the ARL for the Cusum chart with the $h/2$ headstart feature is calculated using the xcusum.arl function.

```
R>#ARL individuals chart with 3 sigma limits
R>mu<-c(0.0,.5,1.0,2.0,3.0,4.0,5.0)
R>OC3=c(pnorm(3-mu[1])-pnorm(-3-mu[1]),
   pnorm(3-mu[2])-pnorm(-3-mu[2]),
   pnorm(3-mu[3])-pnorm(-3-mu[3]),
```

```
    pnorm(3-mu[4])-pnorm(-3-mu[4]),
    pnorm(3-mu[5])-pnorm(-3-mu[5]),
    pnorm(3-mu[6])-pnorm(-3-mu[6]),
    pnorm(3-mu[7])-pnorm(-3-mu[7]))
R>ARL3=1/(1-OC3)
R>#
R>#ARL for Cusum with and without the headstart feature
R>library(spc)
R>ARLC<-sapply(mu,k=.5,h=5,sided="two",xcusum.arl)
R>ARLChs<-sapply(mu,k=.5,h=5,hs=2.5,sided="two",xcusum.arl)
R>round(cbind(mu,ARL3,ARLC,ARLChs),digits=2)
```

To get a vector of Cusum ARLs corresponding to a whole vector of possible mean shift values, the R `sapply` function was used in the code above. This function calculates the ARL for each value of the mean shift by repeating the call to `xcusum.arl` for each value in the vector `mu`. The calculated ARLs form the body of Table 6.4.

TABLE 6.4: ARL for Shewhart and Cusum Control Charts

Shift in Mean Multiples of σ	Shewhart Individuals Chart	Cusum Chart $k = 1/2, h = 5$	Cusum with $h/2$ headstart $k = 1/2, h = 5$
0.0	370.40	465.44	430.39
0.5	155.22	38.00	28.67
1.0	43.89	10.38	6.35
2.0	6.30	4.01	2.36
3.0	2.00	2.57	1.54
4.0	1.19	2.01	1.16
5.0	1.02	1.69	1.02

In this table we can see that the ARL for a Shewhart chart when the mean has not shifted from the known in-control mean is 370 (rounded to the nearest integer), while the ARLs for Cusum charts with or without the headstart feature are longer (i.e. 430–465). Therefore, false positive signals are reduced when using the Cusum chart.

The ARL for detecting a small shift in the mean of the order of 1σ is over 40 for a Shewhart chart plotting individual measurements. Recall from Section 4.5.1, that the ARL for detecting a 1σ shift in the mean using a Shewhart $\overline{X}$ chart with subgroups of size $n = 5$ was approximately 5 (exactly 4.495). That implies $n \times ARL \approx 5 \times 4.49 = 22.45$ total measurements. Therefore, the Shewhart $\overline{X}$ chart can detect small shifts in the mean with fewer measurements than a Shewhart chart plotting individual measurements.

The ARL is only 2.00 for detecting a large shift in the mean of order 3σ using the Shewhart chart plotting individual measurements, while the ARL for detecting the same shift in the mean using a Shewhart $\overline{X}$ chart with subgroups of size $n = 5$ is 1.00 (or $5 \times 1.00 = 5$ total measurements). Thus, the Shewhart plotting individual measurements can detect large shifts in the mean with fewer measurements than the Shewhart $\overline{X}$ chart.

The total number of measurements on average for detecting a small shift in the mean (1σ) with Cusum charts is even less than the Shewhart $\overline{X}$ chart with subgroups of size $n = 5$ (i.e., 10 for the basic Cusum and only 6 for the Cusum with the headstart). Individual measurements can be taken more frequently than subgroups, reducing the time to detect a small shift in the mean using Cusum charts. However, the ARL for detecting a large shift in the mean (3σ–5σ) with the Cusum is greater than Shewhart individuals chart. For this reason a Cusum chart and a Shewhart individuals chart are often monitored simultaneously.

Alternatively, another type of time weighted control chart called the EWMA chart can provide a compromise. It can be constructed so that the ARL for detecting a large shift in the mean (3σ–5σ) will be almost as short as the Shewhart individual chart while the ARL for detecting a small shift in the mean (of the order of 1σ) will be almost as short as the Cusum chart.

6.1.2 EWMA Charts

The EWMA (or exponentially weighted average control chart–sometimes called the geometric moving average control chart) can compromise between the Cusum chart that detects small shifts in the mean quickly and the Shewhart chart that detects large shifts in the mean quickly.

The EWMA control chart uses an exponentially weighted average of past observations, placing the largest weight on the most recent observation. Its value z_n is defined as:

$$z_1 = \lambda x_1 + (1 - \lambda)\mu$$
$$z_2 = \lambda x_2 + (1 - \lambda)z_1$$
$$\vdots$$
$$z_n = \lambda x_n + (1 - \lambda)z_{n-1}$$

Recursively substituting for z_1, z_2 etc, this can be shown to be

$$z_n = \lambda x_n + \lambda(1 - \lambda)x_{n-1} + \lambda(1 - \lambda)^2 x_{n-2} + ... + \lambda(1 - \lambda)^{n-1}x_1 + (1 - \lambda)^n\mu,$$

which equals

$$z_n = \lambda x_n + (1 - \lambda)z_{n-1} = \lambda \sum_{i=0}^{n-1} (1 - \lambda)^i x_{n-i} + (1 - \lambda)^n \mu, \qquad (6.5)$$

where, x_i are the individual values being monitored, μ is the in-control process mean, and $0 < \lambda < 1$ is a constant. With this definition, the weight for the most recent observation is λ, and the weight for the mth previous observation $\lambda(1 - \lambda)^m$, which decreases exponentially as m increases. For a large value of λ, more weight is placed on the most recent observation, and the EWMA chart becomes more like a Shewhart individuals chart. For a small value of λ, the weights are more similar for all observations, and the EWMA becomes more like a Cusum chart.

The control limits for the EWMA chart are given by

$$UCL = \mu + L\sigma\sqrt{\frac{\lambda}{2 - \lambda}\left[1 - (1 - \lambda)^{2i}\right]} \qquad (6.6)$$

$$LCL = \mu - L\sigma\sqrt{\frac{\lambda}{2 - \lambda}\left[1 - (1 - \lambda)^{2i}\right]}, \qquad (6.7)$$

where, μ is the known in-control process mean, σ is the known in-control process standard deviation, and L defines the width of the control limits. L and λ are chosen to control the ARL.

Table 6.5 shows an example of the EWMA calculations using the first 10 simulated observations from the Phase II data shown in Tables 6.2 and 6.1. In this example, $\lambda = 0.2$ and $\mu = 50$ is the known in-control process mean. The first value of the EWMA (z_1) is calculated as $z_1 = 0.2(40.208) + (1 - 0.2)(50) = 48.042$. The next value is calculated as $z_2 = 0.2(56.211) + (1 - 0.2)(48.042) = 49.675$ etc.

Although it is easy to calculate the EWMA (z_i) by hand, the control limits shown in Equations 6.6 and 6.7 are not easily calculated and change with the observation number i.

The function ewma() in the qcc package will compute the EWMA and plot it along with the control limits in Figure 6.4. The R code below illustrates this.

```
R># random data from normal N(50,5) followed by N(56.6, 5)
R>library(qcc)
R>set.seed(109)
R>x1<-rnorm(7,50,5)
R>set.seed(115)
```

```
R>x2<-rnorm(8,56.6,5)
R>x<-c(x1,x2)
R># standardized ewma chart assuming mu=50, sigma=5
R>E1<-ewma(x[1:10],center=50,std.dev=5,lambda=.2,
  nsigmas=2.938)
```

In the function call above, the argument `center=50` is the known in-control mean (μ), `std.dev=5` is the known standard deviation (σ), `lambda=.2` is the value of λ, and `nsigmas=2.938` is the value of L. The choices of $\lambda = 0.2$, and $L = 2.938$ were made to make the ARL for the EWMA Chart, when the mean does not change from $\mu = 50$, equivalent to the ARL for a Cusum chart with $k = 1/2$ and $h = 5$ that was shown in Figure 6.3.

TABLE 6.5: EWMA Calculations $\lambda = 0.2$ Phase II Data from Table 6.3

Sample Number i	Observation x_i	EWMA $z_i = (0.2)x_i + (0.8)z_{i-1}$
1	40.208	48.042
2	56.211	49.675
3	51.236	49.988
4	60.686	52.127
5	45.230	50.748
6	49.849	50.568
7	59.491	50.953
8	59.762	52.715
9	59.462	54.064
10	59.302	55.112

In this chart, the plus signs represent the individual observations x_i, and the connected dots represent the smoothed EWMA values z_i. We see that the EWMA exceeds the upper control limit at the $10th$ observation two observations before the Cusum chart with $k = 1/2$ and $h = 5$. The EWMA is also used as a forecasting formula in business time series applications. So when the EWMA exceeds the control limits, the value of z_i is actually a forecast of the process mean at observation x_{i+1}. Knowing this can help in deciding how to adjust the process to get back to the target mean level μ.

The object list `E1`, created by the `ewma` function in the code above, contains the data in the element `E1$statistics` and the calculated z_i values in the element `E1$y`. The control limits are in the element `E1$limits`. The statements `EW1<-cbind(E1$data,E1$y)` and `colnames(EW1)<-c("x","z")` combine the data and EWMA in a matrix and name the columns. Finally, the statement `cbind(EW1,E1$limits)` creates a matrix containing a Table like Table 6.5 with the addition of the control limits.

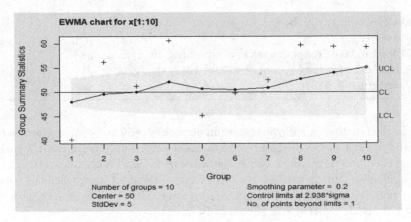

FIGURE 6.4: EWMA for Phase II Data from Table 6.5

6.1.2.1   ARL of EWMA, Shewhart and Cusum Control Charts for Phase II Monitoring

The function xewma.arl in the spc package calculates the ARL for an EWMA chart. For example, to calculate the ARL for an EWMA chart with $\lambda = 0.2$ and $L = 2.938$, as shown in the chart above, use the command:

```
R>library(spc)
R>xewma.arl(mu=0,l=.2,c=2.938,sided="two")
 [1] 465.487
```

The argument mu=0 specifies the shift in the mean in units of σ, the argument c=2.938 specifies L, and sided="two" specifies both upper and lower control limits will be used like those shown in Figure 6.4. The printed result [1] 465.4878 shows the ARL. When there is no shift in the mean, it is equivalent to the ARL of the Cusum chart with $k = 1/2$ and $h = 5$ (as shown in Table 6.6).

Table 6.6 compares the ARL for the Cusum chart with $k = 1/2$ and $h = 5$, the EWMA chart with $\lambda = 0.2$ and $L = 2.938$, the Shewhart individuals chart, and the EWMA chart with $\lambda = 0.4$ and $L = 2.9589$.

TABLE 6.6: ARL for Shewhart and Cusum Control Charts and EWMA Charts

Shift in Mean Multiples of σ	Cusum Chart $k = 1/2$ $h = 5$	EWMA with $\lambda = 0.2$ $L = 2.938$	Shewhart Individuals Chart	EWMA with $\lambda = 0.4$ $L = 2.9589$
0.0	465.44	465.48	370.40	370.37
0.5	38.00	40.36	155.22	58.45
1.0	10.38	10.36	43.89	12.71
2.0	4.01	3.71	6.30	3.35
3.0	2.57	2.36	2.00	1.95
4.0	2.01	1.85	1.19	1.39
5.0	1.69	1.46	1.02	1.10

In this table, it can be seen that the EWMA chart with $\lambda = 0.2$ and $L = 2.938$ is nearly equivalent to the Cusum chart with $k = 1/2$ and $h = 5$. Also, the EWMA chart with a larger value of $\lambda = 0.4$ and $L = 2.9589$ will be almost as quick to detect large $(4-5\sigma)$ shift in the mean as the Shewhart individuals chart. Yet the EWMA ARLs for smaller shifts in the mean are less that $1/3$ that of the Shewhart individuals chart. Clearly, the EWMA control chart is very flexible and can detect both large and small shifts in the mean.

6.1.2.2 EWMA with FIR Feature.

To incorporate a fast initial response feature in an EWMA chart, Steiner[91] suggested incorporating additional constants (f and a) that narrow the time varying control limits of the EWMA for the first few observations following a process restart after a process adjustment. Using these factors the control limits are redefined as:

$$\mu \pm L\sigma \left[\left(1 - (1 - f)^{1+a(t-1)} \right) \sqrt{\frac{\lambda}{2 - \lambda} (1 - (1 - \lambda)^{2t})} \right]. \qquad (6.8)$$

He suggests letting $f = 0.5$, because it mimics the 50% headstart feature usually used with Cusum charts, and $a = [-2/\log(1 - f) - 1]/19$ so that the FIR will have little effect after $t=20$ observations.

Although the `ewma` function in the `qcc` package does not have an option to compute these revised control limits, they could be calculated with R programming statements and stored in the EWMA object element that contains the control limits (like the `E1$limits` discussed above).

6.2   Time Weighted Control Charts of Individuals to Detect Changes in σ

When the capability index (PCR) is equal to 1.00, a 31% to 32% increase in the process standard deviation will produce as much process fallout (or nonconforming output) as a 1σ shift in the process mean. For that reason, it is important to monitor the process standard deviation as well as the mean during Phase II.

Hawkins[36] showed that the distribution of

$$v_i = \frac{\sqrt{|y_i|} - 0.822}{0.349}, \tag{6.9}$$

(where $y_i = (x_i - \mu)/\sigma$) are standardized individual measurements) is close to a standard normal distribution, even when the distribution of the individual values x_i is a non-normal heavy-tailed distribution. Further, he showed that if the process standard deviation increases, the absolute value of y_i increases on the average, and therefore v_i increases.

If the standard deviation of the process characteristic x_i increases by a multiplicative factor γ, then the standard deviation of the standardized value $y_i = (x_i - \mu)/\sigma$ increases to γ, and the expected values of $\sqrt{|y_i|}$ and v_i will correspondingly increase.

If y_i follows a normal distribution with mean zero, and its standard deviation has increased from 1 to γ, the expected value of $\sqrt{|y_i|}$ can be found as shown in Equation 6.10.

$$E[\sqrt{|y_i|}] = \frac{1}{\gamma\sqrt{2\pi}} \int_{\infty}^{\infty} \sqrt{|y_i|} \exp\left(\frac{-y_i^2}{2\gamma^2}\right) dy_i, \tag{6.10}$$

Using numerical integration, the expected values of $\sqrt{|y_i|}$ and v_i for four possible values of γ were found and are shown in Table 6.7.

TABLE 6.7: Expected value of $\sqrt{|y_i|}$ and v_i as function of γ

% Change in σ	γ	$E[\sqrt{|y_i|}]$	$E(v_i)$
−20%	0.8	0.735379	−0.24860
0%	1.0	0.822179	0.00000
32%	1.32	0.944612	0.35066
50%	1.5	1.006960	0.52923

Therefore, a 32% increase in the process standard deviation will result in a 0.35066 standard deviation increase in the mean of v_i, since v_i has mean zero and standard deviation one when the process mean and standard deviation have not changed.

The mean of $\sqrt{|y_i|}$ and v_i not only change when the process standard deviation changes, but they will also change if the process mean (μ) changes. In addition, if the process standard deviation (σ) increases, not only do the means of $\sqrt{|y_i|}$ and v_i increase, but the chance of a false alarm when monitoring the process mean is also more likely.

As pointed out by Hawkins and Olwell[38], this is analogous to the case when using subgrouped data. When the process standard deviation increases, not only does the sample range of subgroups tend to increase, but the chance of a false positive signal being seen on an $\overline{X}$ chart also increases. For that reason, $\overline{X}$ and R charts are usually kept together when they are used for monitoring a process in Phase II.

As already discussed, time-weighted control charts such as the Cusum chart and EWMA chart can detect small shifts in the process mean, during Phase II monitoring, more quickly than Shewhart $\overline{X}$-charts when the process characteristic (x) to be monitored is normally distributed and its "in-control" mean (μ) and standard deviation (σ) are known.

For the same reason that $\overline{X}$ and R charts are usually kept together when they are used for monitoring a process in Phase II, Hawkins[37] recommended keeping a Cusum chart of individual values y_i, together with a Cusum chart of v_i when monitoring individual values in Phase II.

The Cusum chart that Hawkins[36] recommended for monitoring the process standard deviation is defined by Equations 6.11 and 6.12.

$$C_i^+ = max[0, v_i - k + C_{i-1}^+]$$ (6.11)

$$C_i^- = max[0, -k - v_i + C_{i-1}^-],$$ (6.12)

where, $k = .25$ and the decision limit $h = 6$. He showed that if the process standard deviation increases, the Cusum chart for y_i used for monitoring the process mean (shown in column 1 of Table 6.8) may briefly cross its control or decision limit ($h=5$), but that the Cusum for v_i, described above, will cross and stay above its control or decision limit. Therefore, keeping the two charts together will help to distinguish between changes in the process mean and changes in the process standard deviation.

EWMA charts can also be used to monitor changes in individual values like v_i. The EMMA at time point i is defined as:

$$z_i = \lambda v_i + (1 - \lambda)z_{i-1}$$ (6.13)

where $\lambda = 0.05$, and the control limits for the EWMA chart at time point i are given by:

$$LCL = -L\sqrt{\frac{\lambda}{2 - \lambda}[1 - (1 - \lambda)^{2i}]},$$ (6.14)

$$UCL = +L\sqrt{\frac{\lambda}{2-\lambda}\left[1-(1-\lambda)^{2i}\right]}. \qquad (6.15)$$

The `xewma.crit` function in the R package `spc` can be used to determine the value of the multiplier L used in the formula for the control limits so that the ARL_0 of the chart will match that of Hawkins's[36] recommended Cusum chart for monitoring v_i. It is illustrated in the R code shown below where it was used to find the multiplier $L = 2.31934$ needed to produce an $ARL_0 = 250.805$ matching the Cusum chart with $k = .25$ and $h = 6$.

```
R>library(spc)
R>xewma.crit(l=.05,L0=250.8,mu0=0,sided="two")
```

Table 6.8 shows the average run lengths for detecting the positive % change in σ shown in Table 6.7. In this table are shown the ARLs for Hawkins recommended Cusum chart of v_i, the ARLs of the EWMA chart of v_i with similar properties, and the ARLs of a Shewhart individuals chart of v_i with control limits at $\pm 2.88\sigma$ to match the ARL_0 of the other two charts in the table.

TABLE 6.8: ARL for Detecting Increases in σ by Monitoring v_i

% Increase in σ	Cusum Chart $k = .25$ $h = 6$	EWMA with $\lambda = 0.05$ $L = 2.31934$	Shewhart Individuals Chart $\pm 2.88\sigma$ Limits
0%	250.805	250.817	250.8
32%	33.51	33.37	157.56
50%	19.39	19.98	102.93

It can be seen that either the Cusum Chart or EWMA chart of v_i will detect a 32% to 50% increase in the process standard deviation much quicker than the Shewhart individuals chart of v_i. Also, the ARL for detecting a 32% increase in the standard deviation using the standard R chart (with $UCL = d_2(n) \times D_4(n) \times \sigma$, and $LCL = 0$) is 17.44 when the subgroup size is 4, and 15.57 when the subgroup size is 5. This appears to be shorter than the ARL for the Cusum and EWMA charts shown in Table 6.8. However, an ARL 17.44 with a subgroup size of $n = 4$ is $4 \times 17.44 = 69.76$ total observations, and an ARL of 15.57 with a subgroup size of $n = 5$ is $5 \times 15.57 = 77.85$ total observations. This is more than twice the total number of observations required of the Cusum and EWMA charts. In addition, an increase in the process standard deviation could be detected quicker by taking one observation every 15 minutes than it could by taking one subgroup of 4 every hour.

6.3   Examples

As an example of a Cusum chart to monitor process variability with individual values, consider the data shown in the R code below. This data comes from Summers[94] and represents the diameter of spacer holes in surgical tables.

```
R>diameter<-c(.25,.25,.251,.25,.252,.253,.252,
  .255,.259,.261,.249,.250,.250,.250,.252)
R>mu<-.25
R>sigma<-.0025
R>y<-(diameter-mu)/sigma
R>v<-(sqrt(abs(y))-.822179)/.3491508
R>library(qcc)
R>cusum(v,center=0,std.dev=1,decision.interval=6,
  se.shift=.5)
```

There is no known mean or standard deviation for this data, but there are known specification limits of 0.25 ± 0.01. If the mean was $\mu = 0.25$ and the standard deviation was $\sigma = 0.0025$, then $C_p = C_{pk} = 1.33$. A shift of the mean away from $\mu = 0.25$ by more than 0.0025 in either direction would result in $C_{pk} < 1.0$, and an increase in the standard deviation by more than 50% would also result in $C_p < 1.0$. Therefore, although the process mean and standard deviation are unknown, Phase II monitoring of changes in the process mean and standard deviation from the values known to result in $C_p = C_{pk} = 1.33$ could begin, and there is no need for a Phase I study to estimate μ and σ.

In the code above, mu and sigma are assigned the values that would result in $C_p = C_{pk} = 1.33$. The standardized values y are calculated along with the values of v. Next the cusum function in the R package qcc is used to create a Cusum chart of v using value of $k = 0.25$ (i.e. se.shift=.5), and $h = 5$ (i.e. decision.interval=5) to result in an $ARL_0 = 250.805$, as shown in Table 6.8.

Figure 6.5 shows the resulting Cusum chart. It can be seen that the upper Cusum exceeds the upper decision interval at observation 10, but falls back within the limit at observation 12. It is unlikely that the process standard deviation would only temporarily change. To get a better understanding of what has changed in the process, the Cusum chart of the standardized values y was also made using the cusum function call shown in the block of code below.

```
R>library(qcc)
R>cusum(y,center=0,std.dev=1,decision.interval=5,
  se.shift=1)
```

Specifying $h=5$ or decision.interval=5 and $k = .5$ or se.shift=1 will result an ARL_0 of 465.44 (as shown in Table 6.6), and an ARL for detecting

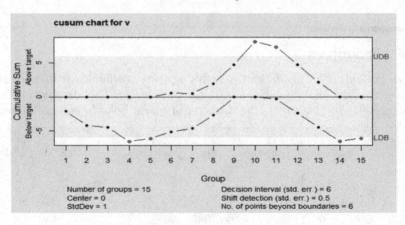

FIGURE 6.5: Cusum Chart of v_i

a one standard deviation change in the mean of 10.38. The resulting Cusum chart is shown in Figure 6.6. There it can be seen that the positive cusums exceeds the upper decision limit at observation 9, and stays above it from that point on. Examination of both of these charts makes it clear that the process mean changed, and the signal on the control chart of v_i was caused by the mean change.

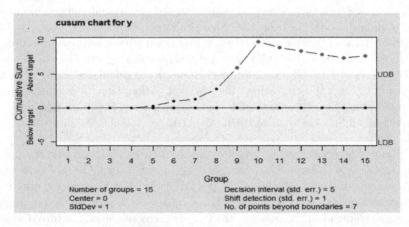

FIGURE 6.6: Cusum Chart of y_i

As a second example of monitoring the process mean and variability with individual observations, consider the block of R code shown below. In this code, the situation where the process standard deviation increases while the process mean remains unchanged is illustrated using randomly generated data. The data is monitored with EWMA charts. 15 observations were generated with a mean of 50 and a standard deviation of 7.5 If the in-control process standard deviation was known to be 5.0, and the in-control mean was known to be 50.0, the random data would represent a 50% increase in the standard deviation with no change in the mean. The emma function in the R package qcc was used to create the two EWMA charts shown in Figure 6.7. The arguments center and std.dev represent the known in-control process mean and standard deviation. The EWMA parameters $\lambda = .2$ and $L = 2.938$ are defined in the first function call with the arguments l=.2 and nsigmas=2.938. The second EWMA (with $\lambda = .05$ and $L = 2.248$) is defined in the second function call with l=.05 and nsigmas=2.248.

```
R>set.seed(29)
R># random data from normal N(50, 7.5)
R>x<-rnorm(15,50,7.5)
R># standardize assuming sigma=5, mu=50
R>y<-(x-50)/5
R># calculate v
R>v<-(sqrt(abs(y))-.822179)/.3491508
R>library(qcc)
R>EWMA <- ewma(y, center=0, std.dev=1, lambda=.2,
          nsigmas=2.938)
R>EWMA <- ewma(v, center=0, std.dev=1,lambda=.05,
          nsigmas=2.248)
```

In Figure 6.7 it can be seen that the EWMA chart of y_i shows no out-of-control signals, while a signal is first shown on the EWMA chart of v_i at observation number two. In this case, since there is no indication that the process mean has changed, the out-of-control signal on the EWMA chart of v_i would be an indication that the process standard deviation has increased.

It is interesting to note that if the 15 data points from these two examples were grouped in three subgroups of size 5, the R chart with UCL=$D_4(5) \times d_2(5) \times \sigma$ would not have shown any out-of-control points. This reiterates the fact that time weighted charts of v_i are more sensitive to small changes in variability, as shown in Tables 6.7 and 6.8.

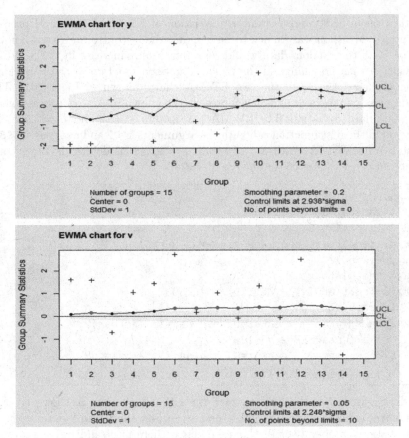

FIGURE 6.7: EWMA Charts of y_i and v_i

6.4 Time Weighted Control Charts Using Phase I estimates of μ and σ

When the the in-control process mean and standard deviation are unknown, the standard recommendation (see Section 6.3.2 of the NIST Engineering Statistics Handbook[1]–Section 6.3.4.3 for CUSUM and 6.3.2.4 for EWMA) is to use the estimates $\hat{\mu} = \overline{\overline{X}}$ and $\hat{\sigma} = \overline{R}/d_2$ from a Phase I study using $\overline{X} - R$ charts with subgroups of size $n = 4$ or 5, and $m = 25$ subgroups. But of course these quantities are random variables.

Figure 6.8 Shows a histogram of 10,000 simulated values of $\overline{R}/d_2\sigma$. Although the distribution is nearly symmetric, it shows that 50% of the time the estimate $\overline{R}/d_2$ will be less than the actual σ and 10% of the time it will be less than 90% of σ.

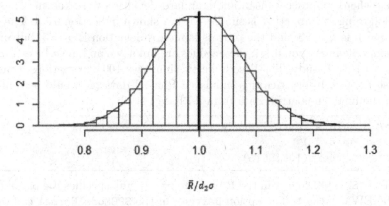

FIGURE 6.8: Simulated Distribution of $\overline{R}/d_2\sigma$

The width of the control limits for the Phase II control charts for monitoring the mean (like the $\overline{X}$ chart or time-weighted charts such as the Cusum or EWMA) are multiples of σ, and when $\overline{R}/d_2$ is used in place of the known σ the control limits will be too narrow at least 50% of the time.

When the control limits are too narrow, the ARL_0 will be decreased and the chance of detecting a spurious out-of-control signal will be increased. This will result in wasted time trying to discover the cause (for which there is none), or if a Phase I OCAP has been established it will result in unnecessary adjustments to the process. The unnecessary adjustments to the process are what Deming[21] has called "Tampering" and will actually result in increased variability in the process output.

In order to prevent the chance of "Tampering" and increasing process variation, the multiplier of σ in the formulas for the control limits should be increased, thus making the control limits wider. If the control limits become wider, the ARL_0 will increase, but the ARL_1 for detecting a change in the process mean will also increase.

Gandy and Kvaloy[30] recommended that a control chart be designed so that the ARL_0 should achieve the desired value with a specified probability. To do this they proposed a method based on bootstrap samples to find the 90th percentile of the control limit multiplier to guarantee the ARL_0 be at least equal to the desired value. They found that the increase in the control limit multiplier only slightly increased the ARL_1 for detecting a shift in the process mean.

The R package spcadjust[29] contains the function SPCproperty() that computes the adjusted multiplier for the control limits of the two-sided EWMA chart based on Gandy and Kvaloy's bootstrap method. It can be used to replace the xewma.crit() function that was illustrated in in Section 6.1.3 to find the multiplier for the control limits to achieve a desired value

ARL_0 when the process mean and standard deviation were known. Consider the following example that is similar to that shown in Section 6.1.4. The process mean is $\mu = 50$, and the process standard deviation is $\sigma = 5$, but unlike the example in Section 6.1.4, these values are unknown and must be estimated from a Phase I study. The R-code below simulates 100 observations from the Phase I study. In practice this simulated Phase I data (X) would be replaced with the final "in-control" data from a Phase I study.

```
R># Simulate Phase I study with 100 observations
R>set.seed(99)
R>X <- rnorm(100,50,5)
```

The "SPCEWMA" class in the R package spcadjust specifies the parameters of the EWMA chart. The option Delta=0 in the SPCModelNormal call below indicates that you want to find the adjusted multiplier that will guarantee the ARL_0 be greater than a desired value with a specified probability.

```
R>library(spcadjust)
R>chart <- new("SPCEWMA",model=SPCModelNormal(Delta=0),
  lambda=0.2);
R>xihat <- xiofdata(chart,X)
R>str(xihat)
```

The xiofdata function computes the Phase I estimated parameters from the simulated data X. In this case $\hat{\mu} = \overline{X}$=xihat$mu=49.5, and $\hat{\sigma} = $ s=xihat$sd=4.5 were calculated from the simulated Phase I data.

Next, the SPCproperty function in the spcadjust package is called to get the adjusted multiplier L for the EWMA chart. Actually this function computes $L \times \sqrt{\frac{\lambda}{2-\lambda}}$ where L is the adjusted multiplier. The option target=465.48 specifies the desired value of ARL_0, and $\lambda = .2$ was specified in "SPCEWMA" class above. The function call is shown below.

```
R>library(spcadjust)
R>cal <- SPCproperty(data=X,nrep=1000,
        property="calARL",chart=chart,
        params=list(target=465.48),quiet=TRUE)
R>cal
90 % CI: A threshold of +/- 1.128 gives an in-control
  ARL of at least 465.48.
Unadjusted result:  0.9795
Based on  1000 bootstrap repetitions.
```

The argument nrep=1000 in the function call specifies 1000 bootstrap samples, which is Gandy and Kvaloy's recommended number to get an accurate estimate of the 90th percentile of the control limit multiplier. The argument property specifies the property to be computed (i.e., the ARL). The argument quiet=TRUE suppresses printing of the progress bar in the R Studio

Console window. After a pause to run the bootstrap samples, `cal` is typed to reveal the value stored in this variable after running the code. It can be seen in the output below the code that $L \times \sqrt{\frac{\lambda}{2-\lambda}} = 1.128$ gives an in-control ARL of at least 465.48. Thus, $L = \frac{1.128}{\sqrt{\frac{.2}{2-.2}}} = 3.384$. This value will change slightly if the code is rerun due to the random nature of the bootstrap samples, but it will guarantee that ARL_0 will be at least 465.48, 90% of the time. Recall from Table 6.6 that if μ and σ are known, $\lambda = .2$ and $L = 2.938$ will result in $ARL_0 = 465.48$.

To illustrate the use of the adjusted multiplier in Phase II monitoring, the code below generates 15 random observations from a normal distribution with mean $\mu = 55$, and standard deviation $\sigma = 5$. This indicates a one standard deviation shift from the unknown but in-control mean. Next the `ewma` function in the `qcc` package is used to make an EWMA chart of the simulated Phase II data. In the `ewma` call, the arguments `center=xihat$mu` and `std.dev=xihat$sd` specify the Phase I estimated values of the mean and standard deviation since the real values are unknown. The arguments `lambda=.2` and `nsigmas=3.384` indicate the value of λ and the adjusted value of L.

```
R># Simulate Phase II data with 1 sigma shift in the mean
R>set.seed(49)
R>x<-rnorm(15,55,5)
R>library(qcc)
R>ewma(x,center=xihat$mu,std.dev=xihat$sd,lambda=.2,
  nsigmas=3.384)
```

Figure 6.9 shows the EWMA chart. There it can be seen that the EWMA first exceeds the upper control limit (indicating an upward shift in the mean) at observation 9. Table 6.6 showed that the ARL_1 for detecting a one standard deviation shift in the mean was 10.36 when $\lambda = 0.2$ and the unadjusted multiplier $L = 2.938$ is used. The fact that the EWMA chart in Figure 6.9 detects the one standard deviation shift in the mean after 9–10 points ($<$ 10.36) illustrates the following statement Gandy and Kvaloy made: Adjusting the multiplier L guarantees the ARL_0 will be at least 465.48, 90% of the time, but will only increase the ARL_1 slightly.

When making an EWMA chart of v_i to monitor the process standard deviation in Phase II, Lawson[61] shows the same strategy can be used to find the adjusted multiplier, L, for the EWMA chart. The R code below shows an example using the same simulated data as shown in the last example. In this case the desired ARL_0 is 250.805 to match the ARL_0 of the Cusum chart (with $k = .25$, and $h = 6$) recommended by Hawkins[36] for monitoring v_i. This code produces the EWMA chart in Figure 6.10.

```
R># Simulate Phase I study with 100 observations
R>set.seed(99)
```

```
R>x1 <-  rnorm(100,50,5)
R># standardize using Phase I estimated mean and
    standard deviation
R>y1<-(x1-mean(x1))/sd(x1)
R>v1<-(sqrt(abs(y1))-.822179)/.3491508
R>library(spcadjust)
R># chart for monitoring sigma using v with lambda=.05
R>chart<-new("SPCEWMA",model=SPCModelNormal(Delta=0),
        lambda=.05)
R># use xiofdata function to get phase I estimates of
    vbar and sd.v
R>xihat<-xiofdata(chart,v1)
R>str(xihat)
R># get the adjusted multiplier for the ewma control
    limits
R>cal<-SPCproperty(data=v1,nrep=1000,property="calARL",
        chart=chart,params=list(target=250.805),
        quiet=TRUE)
R>cal
90 % CI: A threshold of +/- 0.489 gives an in-control
    ARL of at least 250.805.
Unadjusted result:  0.3716
Based on  1000 bootstrap repetitions.
 L=.489/sqrt(.05/(2-.05))=3.0538
R>#Simulate Phase II data with one sigma shift in the mean
  and no shift in the standard deviation
R>set.seed(49)
R>x2<-rnorm(15,55,5)
R>#standardize these simulated values with Phase I
  estimated mean and standard deviation
R>y2<-(x2-mean(x1))/sd(x1)
R>v2<-(sqrt(abs(y2))-.822179)/.3491508
R># make a control chart of the Phase II values of v
    to detect a change from Phase I mean
R>library(qcc)
R>ev<-ewma(v2,center=xihat$mu,std.dev=xihat$sd,
        lambda=.05,nsigmas=3.0538)
```

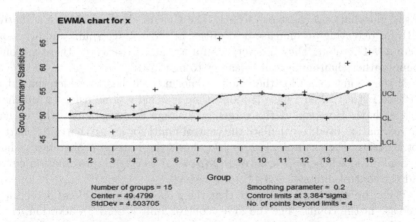

FIGURE 6.9: EWMA chart of Simulated Phase II Data with Adjusted L

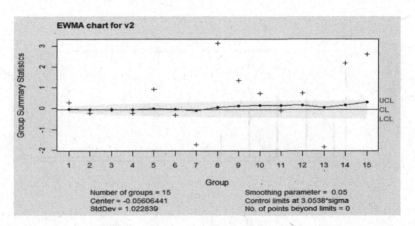

FIGURE 6.10: EWMA chart of v_i from Simulated Phase II Data with Adjusted L

As expected, there are no out of control points on this chart since the standard deviation ($\sigma = 5$) did not change from the Phase I simulated data.

6.5 Time Weighted Charts for Phase II Monitoring of Attribute Data

Cusum charts can be used for counts of the number of nonconforming items per subgroup or the number of nonconformities per inspection unit in place of the Shewhart np chart or c chart. The Cusum charts will have a shorter ARL for detecting an increase or decrease in the average count and are more accurate. Therefore, they are very useful for processes where there are only counts rather than numerical measures to monitor.

If the target mean for the count is known, or it has been determined in a Phase I study, then Phase II monitoring is usually done using an np chart or a c chart. However, if the target or average count is small, the normal approximation used to calculate the control limits for an np chart or c chart is not accurate. For example, Figure 6.11 shows the Poisson probabilities (dark vertical lines) when $\lambda = 2$, a normal ($N(\mu = 2, \sigma = \sqrt{2})$) approximation curve, and the control limits $(2 \pm 3\sqrt{2})$ for a c chart.

It can be seen from the figure that the c chart is incapable of detecting a decrease in the count since the lower control limit is zero. In addition, there is a 0.0166 probability (i.e., false positive rate) of exceeding the upper control limit when the mean count $\lambda = 2$ has not increased. Therefore the average run length to a false positive signal is ARL= $(1/0.0166) = 60.24$.

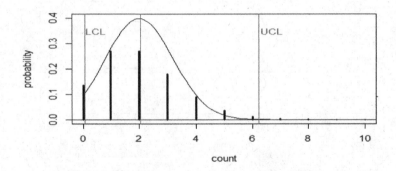

FIGURE 6.11: Normal Approximation to Poisson with $\lambda = 2$ and c chart control limits

6.5.1 Cusum for Attribute Data

Lucas[66] developed a Cusum control scheme for counted data. Of the Cusums shown in Equation 6.16, C_i^+ is used for detecting an increase in D_i, the count of nonconforming items per subgroup (or the number of nonconformities per inspection unit). C_i^- is used for detecting a decrease in the count.

$$C_i^+ = max[0, D_i - k + C_{i-1}^+] \tag{6.16}$$

$$C_i^- = max[0, k - D_i + C_{i-1}^-].$$

He recommended the reference value k be determined by Equation 6.17, rounded to the nearest integer so that the cusum calculations only require integer arithmetic.

$$k = \frac{\mu_d - \mu_a}{\ln(\mu_d/\mu_a)}. \tag{6.17}$$

In this equation, μ_a is an acceptable mean rate (i.e., λ, or np) that is chosen as the Phase I average or a target level, and the Cusum is designed to quickly detect an increase in the mean rate from μ_a to μ_d. The decision interval h is then determined to give a long ARL when there is no shift in the mean and a short ARL when the mean shifts to the undesirable level μ_d.

Lucas[66] provided extensive tables of the ARL (indexed by h and k) for detecting an increase or decrease in the mean count for Cusums with or without a headstart feature at $h/2$. White and Keats[100] developed a computer program to calculate the ARL as a function of k and h. An R function will be shown with the example below that duplicates some features of their program. White, Keats, and Stanley[101] showed that the ARL for this Lucas's Cusum scheme is shorter than a comparable c chart, and that it is robust to the distribution assumption (i.e. Poisson or Binomial).

Consider the application of the Cusum for counted data presented by White, Keats and Stanley[101].

"Motorola's Sensor Products Division is responsible for the production of various lines of pressure sensors and accelerometers. Each of these devices consists of a single silicon die mounted in a plastic package with multiple electrical leads. In one of the final assembly steps in the manufacturing process, the bond pads on the silicon chip are connected to the package leads with very thin (1–2 μ) gold wire in the process known as wire bonding. During this process, the wire is subjected to an electrical charge that welds it to its connections. The amount of energy used in the bond must be carefully controlled because too little energy will produce a bond that is weak and too much energy will severely weaken the wire above the bond and cause subsequent failure of the connection. To maintain control of this process, destructive testing is conducted at regular intervals by selecting a number of units and pulling on the wires until they break. The pull strength as well as the location of the break are recorded. Statistical control of the pull strength falls within the realm of traditional SPC techniques, as these are continuous variable measurements and roughly normal independent and identically distributed. The location of the break is known to engineers as the failure mode. This qualitative variable provides additional information about the state of control for the process. Failure modes for the wire pull destructive testing procedure can take on four values (see Figure 6.12 taken from [White, Keats, and Stanley[101]]). Failure at area 3 is known as a postbond heel break and is a cause of concern if the rate for this failure gets too large. Failure in this area is an indication of an overstressed wire caused by excessive bond energy".

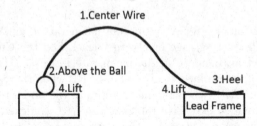

FIGURE 6.12: Wire Pull Failure Modes

The number of post bond heel breaks (D_i) were observed in samples of 16 destructive wire pull tests. If the probability of a the failure mode being the post bond heel break is p, then the D_i follows an Binomial $(16, p)$ distribution, and the mean or expected number would be $16p$. The acceptable mean rate was $\mu_a = 1.88$ (established as either a target or through a Phase I study). An unacceptably high mean rate was defined to be $\mu_d = 3.2$.

If a c-chart was used to monitor the number of post bond heel breaks in Phase II, it would have a center line $\bar{c}$=1.88, with the upper control limit $UCL = \bar{c} + 3 \times \sqrt{\bar{c}} = 5.99$, and lower control limit $= 0$. The average run length (ARL) when there is no change in mean count can be calculated using the ppois function in R as:

$$\mathrm{ARL}_{\lambda=1.88} = 1/(1\text{-ppois}(5,1.88))=79.28,$$

where 5 is the greatest integer less than $UCL = 5.99$. This means a false positive signal would be generated every 80 samples of 16 on the average. The ARL for detecting a shift to $\lambda = \mu_d = 3.2$ with the c chart would be:

$$\mathrm{ARL}_{\lambda=3.2} = 1/(1\text{-ppois}(5,3.2))=9.48.$$

If Lucas's Cusum control scheme for counted data was used in Phase II instead of the c-chart, k would be calculated to be:

$$k = \frac{\mu_d - \mu_a}{\ln(\mu_d/\mu_a)} = \frac{3.2 - 1.88}{ln(3.2) - ln(1.88)} = 2.48, \tag{6.18}$$

which was rounded to the nearest integer $k = 2.0$.

With $k = 2.0$, the next step is to choose h. Larger values of h will result in longer ARLs both when $\lambda = \mu_a = 1.88$, and when $\lambda = \mu_d = 3.2$.

The function arl in the in the R package IAcsSPCR (that contains the data and functions from this book) was used to calculate $\mathrm{ARL}_{\lambda=1.88}$, and $\mathrm{ARL}_{\lambda=3.2}$ for the Cusum control scheme for counted data with k held constant at 2.0, and various values for h. The values of h chosen were 6, 8, 10, and 12. By choosing even number integers for h, $h/2$ is also an integer and the cusum calculations with or without the FIR feature will only involve integer arithmetic. The function calls in the block of code below were used to calculate the entries in Table 6.9.

```
R> library(IAcsSPCR)
R> arl(h=6,k=2,lambda=1.88,shift=0)
R> arl(h=6,k=2,lambda=1.88,shift=.9627)
R> arl(h=8,k=2,lambda=1.88,shift=0)
R> arl(h=8,k=2,lambda=1.88,shift=.9627)
R> arl(h=10,k=2,lambda=1.88,shift=0)
R> arl(h=10,k=2,lambda=1.88,shift=.9627)
R> arl(h=12,k=2,lambda=1.88,shift=0)
R> arl(h=12,k=2,lambda=1.88,shift=.9627)
```

For example, when $h = 6$ and $k = 2$ for the case where λ has increased to 3.2, the ARL is calculated with the function call `arl(h=6,k=2,lambda=1.88,shift=.9627)`, since $3.2 = \lambda = 1.88 + .9627 \times \sqrt{1.88}$, or an increase in the mean by 0.9627 standard deviations. The resulting ARLs for $h = 6$, 8, 10, and 12 are shown in Table 6.9.

TABLE 6.9: ARL for Various Values of h with $k = 2$, $\lambda_0 = 1.88$

Value of h	ARL$_{\lambda=1.88}$	ARL$_{\lambda=3.2}$
6	37.20	5.49
8	66.52	7.16
10	108.60	8.82
12	166.98	10.49

From this table, it can be seen that when h is 10, the ARL$_{\lambda=1.88} = 108.60$ which is 37% longer than the ARL$_{\lambda=1.88} = 79.28$ that would result if a c chart were used. Additionally, the ARL$_{\lambda=3.2} = 8.82$ is 7% shorter than the ARL$_{\lambda=3.2} = 9.48$ that would result for the c chart. This is the only case where both ARL$_{\lambda=1.88}$ and ARL$_{\lambda=3.2}$ for the Cusum scheme for counted data differ from the ARLs for the c chart in the desired directions. Therefore, this is the value of h that should be used.

The calculated cusums (using C_i^+ in Equation 6.16 for the number of the number of post bond heel breaks (D_i) on machine number two in the article by White, Keats and Stanley[101] is shown in Table 6.10. Here it can be seen that the Cusum without the FIR detects a positive signal at sample number 10, where $C_{10}^+ > h = 10$. This would indicate that the number of heel breaks has increased, and that the amount of energy used in the bonding process should be reduced to bring the process back to the target level.

The Cusum including the FIR at $(C_0^+ = h/2 = 5)$ detects a positive signal even sooner at sample number 8. Use of this Cusum would be appropriate when restarting the process after a manual adjustment of the energy level. The c chart shown in the article did not detect an increase until sample number 14. Therefore, the Cusum chart detected the increase in the mean faster and had a 37% increase in the expected average time between false positive signals. Also, with an integers for k and $h/2$, the arithmetic needed to calculate C_i^+ is so simple it can be done in your head as you write the last three elements of each row in a table like 6.10.

TABLE 6.10: Calculations for Counted Data Cusum for Detecting an Increase with $k = 2$, $h = 10$

Sample No. i	D_i	$D_i - k$	C_i^+ with No FIR	C_i^+ with FIR $h/2$
1	3	1	1	6
2	1	-1	0	5
3	4	2	2	7
4	1	-1	1	6
5	3	1	2	7
6	1	-1	1	6
7	5	3	4	9
8	4	2	6	11
9	5	3	9	-
10	5	3	12	-

R code can also be used to calculate the entries in Table 6.10 as shown in the example below. This block of code also illustrates how to produce the graphical representation of the Cusum chart without and with the headstart feature shown in Figure 6.13.

```
R>#Calculations to Get Table 6.10
R>library(qcc)
R>D<-c(3,1,4,1,3,1,5,4,5,5)
R>k<-2
R>CUSUM<-cusum(D,center=0,std.dev=1,decision.interval=10,
        se.shift=4,plot=FALSE)
R>FIR<-cusum(D,center=0,std.dev=1,decision.interval=10,
        se.shift=4,head.start=5,plot=FALSE)
R>M<-cbind(D,(D-k),CUSUM$pos,FIR$pos)
R>colnames(M)<-c("D","D-k","C+(without(FIR))",
        "C+(with(FIR))")
R>M
```

By specifying `center=0` and `std.dev=1`, in the calls to the `cusum()` function, the values of C_i^+ shown in Equation 6.16 without the FIR feature can be found in the named element CUSUM$pos of the object CUSUM, and the vector of values for C_i^+ with the FIR feature can be found in the named element FIR$pos of the object FIR. $C_0^+ = 0.0$ for the Cusum without FIR, and $C_0^+ = 5$ for the Cusum with FIR. The number of postbond heel breaks is stored in the vector D.

When monitoring to detect an improvement or reduction in the Poisson process mean, μ_d will be less than μ_a, and in that case the values the values of $-C_i^-$ the negatives of C_i^- shown in Equation 6.16 can be found in the element CUSUM$neg when the call to the cusum() function is

```
R>CUSUM<-cusum(D,center=0,std.dev=1,decision.interval=10,
        se.shift=-6)
```

and -6 in the argument shift=-6 is the negative of k found using Equation 6.18. When the objective is to detect increases in the Poisson process mean, the line plot on the negative side of the cusum chart plotted by the cusum function should be ignored, and when the objective is to detect a decrease in the Poisson mean by assigning the shift argument to be the negative of k, the line graph graph on the positive side of the cusum chart created by the cusum function should be ignored.

The advantage for using a Cusum for counted data rather than a c chart or np chart for Phase II monitoring is usually much greater than that shown in the last example. To illustrate, consider the case where the target or Phase I in-control average count was $\lambda = 4$. In this case, Table 6.11 compares the expected ARLs for the c chart and the Cusum chart with and without the FIR feature.

TABLE 6.11: Comparison of ARL for c chart and Cusum for Counted Data ($k = 5$, $h = 10$) with Target $\lambda = 4$

λ	ARL for c chart	ARL for Cusum without FIR	ARL for Cusum with FIR
4	352.14	421.60	397.5
5	73.01	29.81	22.38
6	23.46	9.73	6.11
7	10.15	5.59	3.35
10	2.40	2.58	1.58

In this table, it can be seen that there is a 12.9% to 19.7% increase in the average time between false positive signals (depending on whether the FIR feature is used or not). Further, there is a 58% to 74% decrease in the average time to detect a one standard deviation increase in the average count (from $\lambda = 4$ to $\lambda = 6$).

Although a Cusum control chart will detect a small shift in the mean quicker than a Shewhart control chart, it will not detect a large shift in the

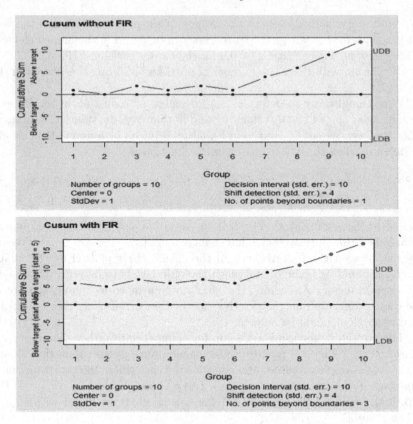

FIGURE 6.13: Cusum for counts without and with FIR

mean as quickly as a Shewhart chart. For example, Table 6.4 shows that the Cusum chart will detect a 0.5–1.0 standard deviation shift in the mean quicker than the Shewhart individuals chart, but the Shewhart chart will detect a 4.0–5.0 shift in the mean quicker than the Cusum chart. Likewise, Table 6.11 shows that a Cusum chart for counted data will detect a 0.5 to 1.5 standard deviation shift in the mean (i.e., $\lambda = 4$ to $\lambda = 5, 6$ or 7) faster than the Shewhart c chart, but that the c chart will detect a 3.0 standard deviation shift (i.e. $\lambda = 4$ to $\lambda = 10$) faster than the Cusum chart.

This is because the Cusum chart weights all past observations leading up to the out-of-control signal equally, while the Shewhart chart ignores all but the last observation. If the mean suddenly shifted by a large amount, the Cusum value (C_i^+ or C_i^-) de-weights the last observation by averaging it with the past. Therefore, a Shewhart individuals chart and Cusum chart are often used simultaneously in Phase II monitoring.

By using the FIR feature, the Cusum chart can be made to detect a large shift as quickly as a Shewhart chart. However, this feature is normally only used after the process has been manually reset after an out-of-control signal. Always using the FIR feature can lead to a reduction in the ARL to encountering a false positive.

6.5.2   EWMA for Attribute Data

The EWMA chart like the Cusum chart can also be used for attribute data like counts of the number of nonconforming items per subgroup or the number of nonconformities in an inspection unit. The EWMA ($z_i = \lambda x_i + (1 - \lambda)z_{i-1}$, where $z_0 = \mu_0$) is a weighted average of past observations or counts (x_i), and due to the Central Limit Theorem it will be approximately normally distributed.

Borror, Champ and Rigdon[7] present the control limits for an EWMA chart for counts as shown in Equations 6.19 and 6.20.

$$UCL = \mu_0 + A\sqrt{\frac{\lambda\mu_0}{2 - \lambda}} \qquad (6.19)$$

$$LCL = \mu_0 - A\sqrt{\frac{\lambda\mu_0}{2 - \lambda}} \qquad (6.20)$$

Where, A is the control limit factor, and μ_0 is the target count (or mean established in Phase I). Once μ_0 is known and λ is selected, the factor A should be selected to control the ARL for various values of the shift in the mean. The paper gives graphs of the ARL as functions of λ, and A that can be used to do this. There is an example of one of these graphs in the lecture slides for Chapter 6. The paper also shows that the EWMA control chart for counts will detect a small shift in the average count faster than the Shewhart c–chart.

6.6   Exercises

1. Run all the code examples in the chapter, as you read.

2. Assuming the target mean is 15 and the in-control standard deviation is $\sigma = 2.0$,

    (a) Make a cusum chart with h = 4 that will quickly detect a one σ shift in the mean in the following Phase II data: 10.5 14.0 15.4 10.7 13.7 13.3 16.4 17.2 14.2 12.6 12.9 13.4 13.0 10.7 10.8 12.9 12.1 10.1 11.4 10.9

    (b) Calculate the ARL_0 for the chart you constructed in part (a).

3. If $\lambda = .3$, find the value of L for an EWMA chart that would match the ARL_0 of the Cusum chart you used in question 2.

    (a) Make the EWMA chart of the data in exercise 2 using $\lambda = .3$ and the value of L you determined above.

    (b) Are the results the same as the Cusum chart in exercise 2?

4. E.L. Grant(1946) showed the following data from a Chemical plant. The numbers represent the daily analyses of the percentages of unreacted lime (CaO) at an intermediate stage of a continuous process.

    .24,.13,.11,.19,.16,.17,.13,.17,.10,.14,.16,.14,.17,.15,.20,.26,.16,0.0,.18,

    .18,.20,.11,.30,.21,11,.17,.18,.13,.28,.16,.14,.16,.14,.10,.13,.20,.14,.10,

    .18,.11,.08,.12,.13,.12,.17,.10,.09

    (a) Make a Cusum chart of the data assuming the in-control mean was $\mu = 0.15$ and the in-control standard deviation $\sigma = .04$. Use $k = \frac{1}{2}$ and $h = 5$.

    (b) Remake the chart using the headstart feature. Is there any difference in the two charts?

    (c) When would you recommend using the headstart feature with individual data from a chemical plant?

    (d) Make a EWMA chart with the similar ARL_0 and the headstart feature.

    (e) Is the result the same as the Cusum chart used in (b)

5. Using Borror et. al.(1998)'s method, make an EWMA control chart of the count data data in Table 6.10 assuming $\mu_0 = 1.88$ and using $\lambda = 0.2$, and $A = 2.55$.

(a) Do you detect any out of control signal? If so at what observation number?

(b) If $\mu_0 = 4.0$, and $\lambda = 0.3$, what values of A would you recommend to obtain an $ARL_0 = 500$ for an EWMA chart made with the Borror et. al.(1998) method.

6. Hawkins proposed a Cusum chart of $v_i = \frac{\sqrt{|y_i|}-0.822}{0.349}$, where $y_i = (x_i - \mu)/\sigma$ to detect changes in the standard deviation of x_i.

(a) Can you use an EWMA chart with individual observations to detect an increase in the standard deviation?

(b) If so, do it with the data in the first column of Table 6.1 and assuming $\mu_0 = 50$ and $\sigma_0 = 5$.

7. The data in the data vector x2 in the R package IAcsSPCR is from Phase II monitoring. The in-control mean and standard deviation are unknown, but the data in the vector x1 in the R package IAcsSPCR are from a Phase I study after eliminating out-of control data points.

(a) Find the 90th percentile of the multiplier L for the control limits of a Phase II EWMA chart with $\lambda = .2$ so that the ARL_0 will be guaranteed to be at least 465.48 90% of the time.

(b) Use the multiplier you found in (a) to construct the EWMA control chart of the data in the vector x2.

(c) Has the process remained in control during Phase II?

7

Multivariate Control Charts

7.1 Introduction

When the quality of a product or service is defined by more than one property, all the properties should be studied simultaneously to control and improve quality (Kourti and MacGregor[56]). The rapid growth of online data acquisition makes it possible to collect and study many properties, and there is a need to use control charts to monitor all the properties, or quality characteristics, simultaneously.

Examples of process outputs that have multiple quality characteristics are as follows. First, in the output of a chemical manufacturing process, the presence of several impurities indicates a lack of quality. These impurities can be identified as peaks on a chromatography report, and all should be reduced to improve quality. Another example is in printed circuit board manufacturing, in which there are many variables measured on each board that defines its quality. Third, in healthcare, many simultaneously measured outcomes contribute to the quality of a service to patients. Finally, a manufactured part may have several physical dimensions that can be measured, and they must all be within a joint region to insure the part fitness for use.

If there are p quality characteristics and separate control charts that are maintained on each with $\pm 3\sigma$ control limits, the probability of a false signal from any one control chart is 0.0027 and the $ARL_0 = 370$. However, the probability of a false signal from at least one of the p control charts is increased to $1 - (1 - .0027)^p$, and the overall ARL_0 will be greatly decreased, resulting in frequent false-positive signals if the p quality characteristics being monitored are independent.

One way to compensate for this problem is to widen the control limits on each of the separate control charts, thus increasing each of their individual ARL_0's to the point that the overall ARL_0 for obtaining a false signal on at least one of the charts is still equal to 370. Again, that can be done if the quality characteristics are jointly independent or uncorrelated.

However, if the several properties or quality characteristics measured on one product or service are interrelated or correlated, the opposite problem may occur. The chance of missing a real change in the process is increased, and although several charts are maintained, together they may be insensitive to detecting real changes in the process. Widening the limits of each individual control chart in this situation makes the charts even less sensitive to detecting real changes.

For example, Figure 7.1 illustrates two interrelated quality characteristics X_1 and X_2. In this figure X_1 and X_2 are negatively correlated. Independent 99% tolerance regions are represented for both X_1 and X_2 as the lengths of the horizontal and vertical lines, and the joint 99% tolerance region for X_1 and X_2 is represented by the ellipse. It can be seen that an observation on $(X_1$ and $X_2)$ could be very unusual and outside the tolerance ellipse, yet well within the independent tolerance regions for X_1 and X_2.

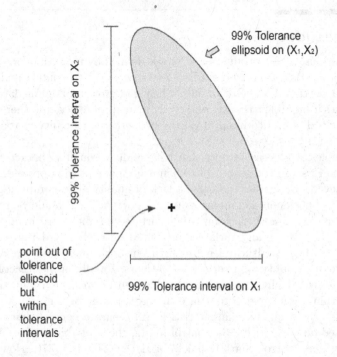

FIGURE 7.1: Comparison of Elliptical and Independent Control

Thus, when several quality characteristics are used in either a Phase I study to establish the characteristics of an in-control process and to develop an OCAP, or during Phase II monitoring, the following considerations are important (Bersimis et. al.[6]):

1. Reduce the chance of false-positive and false-negative signals when determining whether the process is in control.

2. Take into account any correlation among the quality characteristics studied.

3. If the process is out of control, be able to identify the nature of the problem.

This chapter will address these concerns using multivariate control charts. There are four distinct situations where multivariate control charts are used:

1. Phase I with data in rational subgroups

2. Phase I with individual observations

3. Phase II with data in rational subgroups

4. Phase II with individual observations.

All four of these situations will be discussed in this chapter.

7.2 T^2-Control Charts and Upper Control Limit for T^2 Charts

If a single quality characteristic X follows a normal distribution, the statistic for testing the hypothesis $H_0 : \mu = \mu_0$ when σ is known is: $Z = \frac{\bar{X}-\mu_0}{\sigma/\sqrt{n}}$, and using rational subgroups of size n, a standardized control chart for the mean is made by charting the quantity $Z_i = \frac{\bar{X}_i-\mu_0}{\sigma/\sqrt{n}}$, where $\bar{X}_i$ is the ith subgroup mean, μ_0 is the known in-control process mean, and σ is the known in-control process standard deviation. The control limits are $\pm Z_{\alpha/2}$, where $Z_{\alpha/2}$ is the $\left(\frac{\alpha}{2}\right)$th quantile of the standard normal distribution.

When there are p correlated quality characteristics that follow a multivariate normal distribution, the statistic for testing the hypothesis $H_0 : \mu = \mu_0$,

where $\mu_0 = \begin{pmatrix} \mu_{01} \\ \mu_{02} \\ \vdots \\ \mu_{0p} \end{pmatrix}$ is a hypothesized mean vector, and Σ is the known

covariance matrix, is: $T^2 = n(\bar{x} - \mu_0)'\Sigma^{-1}(\bar{x} - \mu_0)$, and $\bar{x} = \begin{pmatrix} \bar{x}_1 \\ \bar{x}_2 \\ \vdots \\ \bar{x}_p \end{pmatrix}$ is a

sample mean vector of n vectors of p quality characteristics. The distribution of this statistic is χ^2 with p degrees of freedom.

Therefore, if rational subgroups of size n are collected, a standardized control chart for the mean vector is made by charting the quantities

$$T_i^2 = n(\bar{\mathbf{x}}_i - \mu_0)'\Sigma^{-1}(\bar{\mathbf{x}}_i - \mu_0) \tag{7.1}$$

where $\bar{x}_i = \begin{pmatrix} \bar{x}_1 \\ \bar{x}_2 \\ \vdots \\ \bar{x}_p \end{pmatrix}$ is the ith subgroup mean vector, $\mu_0 = \begin{pmatrix} \mu_{01} \\ \mu_{02} \\ \vdots \\ \mu_{0p} \end{pmatrix}$ is the

known in-control process mean vector, and Σ is the known in-control process covariance matrix. The lower control limit for this chart is zero, and the upper control limit is $\chi^2_{\alpha/2,p}$. Plotting the quantities in Equation 7.1 along with the upper control limit is called the multivariate T^2 control chart.

When the in-control mean $\mu_0 = \begin{pmatrix} \mu_{01} \\ \mu_{02} \\ \vdots \\ \mu_{0p} \end{pmatrix}$ and in-control covariance ma-

trix Σ are unknown, they must be estimated by $\bar{\bar{x}}$ and S from a series of in-control points on a Phase I control chart. If a Phase I study used m sub-

groups of size n, $\bar{\bar{x}} = \begin{pmatrix} \bar{\bar{x}}_1 \\ \bar{\bar{x}}_2 \\ \vdots \\ \bar{\bar{x}}_p \end{pmatrix}$, and

$$S = \begin{bmatrix} \bar{s}^2_j & \cdots & \bar{s}_{1p} \\ & \ddots & \\ \bar{s}_{p1} & \cdots & \bar{s}^2_p \end{bmatrix},$$

where $\bar{\bar{x}}_j$ is the mean over the m subgroups of the subgroup sample means ($\bar{x}_j = \sum_{l=1}^{n} x_{jl}/n$)) for the jth quality characteristic, $\bar{s}^2_j$ is the mean over the m subgroups of the sample variances ($s^2_j = \sum_{l=1}^{n} (x_{jl} - \bar{x}_l)^2/(n-1)$ for quality characteristic j within each subgroup, and $\bar{s}_{jk}$ is the mean over the m subgroups of the sample covariances $s_{jk} = \sum_{l=1}^{n} (x_{jl} - \bar{x}_j)(x_{kl} - \bar{x}_k)/(n-1)$ between quality characteristics j and k within each subgroup.

If the Phase I control chart used individual samples, then $\bar{x}$ is the mean vector of the p quality characteristics over the m individual values, and S is the sample covariance matrix over the m individual pairs of the p quality characteristics.

The control limits for a Shewhart control chart are normally set at the mean plus or minus three standard deviations. When the process mean and standard deviation are unknown and must be estimated from a Phase I study, the control limit multiplier should be increased to prevent false positive signals. In Chapter 6, section 6.2, the SPCproperty() function in the spcadjust package was used to get larger control limit multipliers for the EWMA charts used in Phase II. The χ^2_p limit for a T^2 control chart used when the process mean and covariance matrix are known is similar to the ± 3 control limit multipliers for Shewhart charts. However, the upper control limit should again be increased when the process mean and covariance matrix are estimated from a Phase I study.

When the data on p correlated quality characteristics is gathered in m subgroups of size n, the ith T^2 statistic plotted on the control chart is:

$$T_i^2 = n(\bar{\mathbf{x}}_\mathbf{i} - \bar{\bar{\mathbf{x}}})' \mathbf{S}^{-1} (\bar{\mathbf{x}}_\mathbf{i} - \bar{\bar{\mathbf{x}}}), \tag{7.2}$$

and the upper control limit for a Phase I control chart changes from $UCL = \chi^2_{\alpha,p}$ to:

$$UCL = \frac{p(m-1)(n-1)}{mn - m - p + 1} F_{\alpha,p,mn-m-p+1}. \tag{7.3}$$

For Phase II monitoring the T^2 statistic is

$$T_f^2 = n(\bar{\mathbf{x}}_\mathbf{f} - \bar{\bar{\mathbf{x}}})' \mathbf{S}^{-1} (\bar{\mathbf{x}}_\mathbf{f} - \bar{\bar{\mathbf{x}}}), \tag{7.4}$$

where $\bar{\mathbf{x}}_\mathbf{f}$ is a future subgroup mean to be observed, and the control limit changes to:

$$UCL = \frac{p(m+1)(n-1)}{mn - m - p + 1} F_{\alpha,p,mn-m-p+1}. \tag{7.5}$$

where m is the number of subgroups of size n used in the Phase I study to estimate the process mean vector and covariance matrix.

When the data on p correlated quality characteristics is gathered in a Phase I study with m individual observations, the ith T^2 statistic plotted on the control chart is:

$$T_i^2 = (\mathbf{x}_\mathbf{i} - \bar{\mathbf{x}})' \mathbf{S}^{-1} (\mathbf{x}_\mathbf{i} - \bar{\mathbf{x}}), \tag{7.6}$$

$\bar{\mathbf{x}}$, and $\mathbf{S}$ are the sample mean vector and covariance matrix and the upper control limit for a Phase I control chart is:

$$UCL = \frac{(m-1)^2}{m} \beta_{\alpha,p/2,(m-p-1)/2}, \tag{7.7}$$

where $\beta_{\alpha,p/2,(m-p-1)/2}$ is the α percentile of the beta distribution with parameters $p/2$ and $(m - p - 1)/2$. The upper control limit for the Phase II control limit is:

$$UCL = \frac{p(m+1)(m-1)}{m^2 - mp} F_{\alpha,p,mn-p}, \tag{7.8}$$

again where m is the number of subgroups of size 1 used in the Phase I study to estimate the process mean vector and covariance matrix.

$$\lambda = (\mu - \mu_0)' \Sigma^{-1} (\mu - \mu_0), \tag{7.9}$$

is the noncentrality parameter for detecting a shift in the mean from μ_0. to μ.

7.3 Multivariate Control Charts with Sub-grouped Data

7.3.1 Phase I T^2 Control Chart with Sub-grouped Data

Table 9.2 in Ryan[83] presents data for $m=20$ subgroups of $n=4$ observations on $p=2$ quality characteristics from a Phase I study. That book also illustrates the use of Minitab®, a commercial program, to construct a Phase I T^2 control chart. The data is included as the dataframe `RyanMultivar` in the R package qcc. That dataframe is in the format for the `mqcc()` function in the qcc package that makes, T^2 control charts. However, it is not in the format that would normally be used to store multivariate data. The `IAcsSPCR` package, that contains the dataframes from this book, also contains the same data (`Ryan92`) in a more familiar format. This format includes one line for each observation and one column for each quality characteristic. The data is in time order sequence with a subgroup indicator variable in the first column. The R code below illustrates retrieving the dataframe and reformatting it so that the `mqcc()` function in the qcc package can be used to create a T^2 chart. The first 10 lines of the dataframe are printed with the `head()` function. It can be seen that the first column is the subgroup number, the second and third columns are the values of the $p=2$ quality characteristics x1 and x1.

```
R>library(IAcsSPCR)
R>data(Ryan92)
R>head(Ryan92,10)
   subgroup x1 x2
1         1 72 23
2         1 84 30
3         1 79 28
4         1 49 10
5         2 56 14
6         2 87 31
7         2 33  8
8         2 42  9
9         3 55 13
10        3 73 22
R># reformat as a list of matricies required by the mqcc
    function
R>X1<-matrix(Ryan92$x1,nrow=20,byrow=TRUE)
R>X2<-matrix(Ryan92$x2,nrow=20,byrow=TRUE)
R>XR = list(X1 = X1, X2 = X2) # a list of matrices,
  one for each variable
```

The mqcc() function requires the data be formatted as a list of m by n matrices one for each quality characteristic. The next two statements in the code above extract the quality characteristic columns from the data frame, and then reformat each of them into a $m = 20 \times 4 = n$ matrices. Finally these matrices are combined in a list named XR.

The next block of code illustrates the use of the mqcc() function to create the T^2 chart shown in Figure 7.2. The summary of the object q created by the mqcc() function is shown below the code.

```
R>library(qcc)
R>q = mqcc(XR, type = "T2",add.stats=TRUE,
   title="T2 chart for Data in Ryan's Table 9.2")
R>summary(q)


-- Multivariate Quality Control Chart ------------

Chart type                = T2
Data (phase I)            = X
Number of groups          = 20
Group sample size         = 4
Center =
      X1        X2
60.3750  18.4875
Covariance matrix =
         X1            X2
X1 222.0333  103.11667
X2 103.1167   56.57917
|S| = 1929.414


Control limits:
 LCL        UCL
   0  11.03976
```

in the summary output it can be seen that the estimated mean vector $\bar{\bar{\mathbf{x}}} = \begin{pmatrix} 60.3750 \\ 18.4875 \end{pmatrix}$, and the estimated covariance matrix

$$S = \begin{bmatrix} 222.0333 & 103.11667 \\ 103.11667 & 56.57917 \end{bmatrix}.$$

The UCL=11.03976. The mqcc() function uses $\alpha = 1 - (1 - .0027)^p$ in order to reduce the chance of false positive results. Knowing this value of α the UCL can be verified using Equation 7.3.

UCL= $\frac{p(m-1)(n-1)}{mn-m-p+1} F_{\alpha,p,mn-m-p+1}$ as shown in the code below.

```
R>m<-20
R>p<-2
R>n<-4
R>num<-m*n*p-m*p-n*p+p
R>dfd<-m*n-m-p+1
R>alpha<-1-(1-.0027)^p
R>UCL<-(num/dfd)*qf(1-alpha,p,dfd)
R>UCL
[1] 11.03976
```

The T^2 chart in Figure 7.2 shows subgroups 10 and 20 to be out of control. Since the sample covariance matrix **S** is not diagonal, the two quality characteristics x1 and x2 are not independent and we would not expect $\bar{X}$-charts constructed for each quality characteristic to be as sensitive to detecting changes as the T^2 chart. The code below makes the two $\bar{X}$-charts that assume independence. Running this code (left as an exercise) shows that subgroups 10 and 20 are within the control limits on both individual control charts. The last statement in the code produces a plot similar to Figure 7.1 that identifies why subgroups 10 and 20 are unusual. The plot it creates is shown in Figure 7.3.

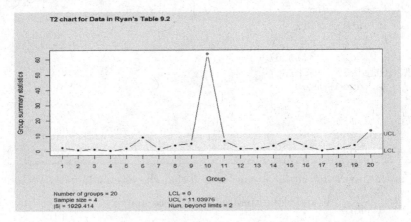

FIGURE 7.2: Phase I T^2 Chart

```
R>qcc(X1,type="xbar",sizes=4)
R>qcc(X2,type="xbar",sizes=4)
R>ellipseChart(q, show.id = TRUE)
```

In Figure 7.3, it can be seen that although the x1 coordinate and the x2 coordinate for subgroups 10 and 20 are not unusual on their own, the two pairs

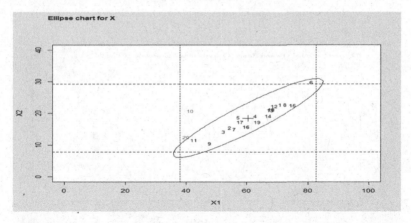

FIGURE 7.3: Ellipse Plot Identifying Out-of-control Subgroups

of coordinates lie outside the control ellipse. Normally there is a strong positive correlation or relationship between characteristic x1 and x2, but subgroups 10 and 20 are different.

The cause for these two out-of-control subgroups should be investigated. Any causes and corrective actions found should be added to the out-of-control action plan (OCAP) as described in Chapter 4. The control chart should be reconstructed without the out-of-control points to get better estimates of the in-control mean vector and covariance matrix. This is illustrated in the R code below.

```
R>library(IAcsSPCR)
R>data(Ryan92)
R># the following statement eliminates subgroup 10
R>Ryan92s<-subset(Ryan92, subgroup != 10)
R># the following statement eliminates subgroup 20
R>Ryan92s<-subset(Ryan92s, subgroup != 20)
R>X1<-matrix(Ryan92s$x1,nrow=18,byrow=TRUE)
R>X2<-matrix(Ryan92s$x2,nrow=18,byrow=TRUE)
R>XR2 = list(X1 = X1, X2 = X2) # a list of matrices,
   one for each variable
R>library(qcc)
R>q2 = mqcc(XR2, type = "T2",add.stats=TRUE,
       title="T2 chart for Data in Ryan's Table 9.2
       eliminating Subgroups 10 and 20")
R>summary(q2)
```

The resulting T^2 chart is shown in Figure 7.4.

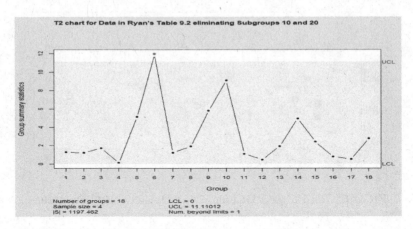

FIGURE 7.4: Phase I T^2 Chart Eliminating Subgroups 10 and 20

Now it can be seen that subgroup 6 is beyond the UCL. The process of refining the control limits and adding items to the OCAP is often an iterative process as described in Chapter 4. The next step is to search for the cause for subgroup 6 falling out of the limits. The ellipse plot, created using the statement `ellipseChart(q, show.id = TRUE)` as illustrated above, may spark ideas in those familiar with the process as to the cause for the unusual subgroup. The result in Figure 7.5 shows subgroup six is slightly above and to the right of the majority of points.

The code below was used to re-create the T^2 chart eliminating subgroup 6. The output of the summary, below the code, shows the revised estimates of the mean vector, covariance matrix, and the revised UCL.

```
R># the next statment eliminates subgroup 6
R>Ryan92s<-subset(Ryan92s, subgroup != 6)
R>X1<-matrix(Ryan92s$x1,nrow=17,byrow=TRUE)
R>X2<-matrix(Ryan92s$x2,nrow=17,byrow=TRUE)
R>XR3 = list(X1 = X1, X2 = X2) # a list of matrices, one for
    each variable
R>library(qcc)
R>q3 = mqcc(XR3, type = "T2",add.stats=TRUE,
R>  title="T2 chart for Data in Ryan's Table 9.2 Eliminating
    Subgroups, 6, 10, and 20")
R>summary(q3)
-- Multivariate Quality Control Chart ------------

Chart type               = T2
```

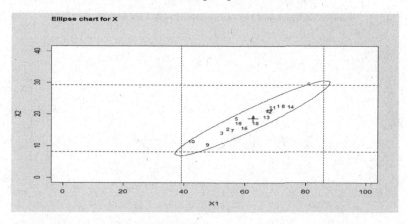

FIGURE 7.5: Ellipse Plot Identifying Out-of-control Subgroup 6

```
Data (phase I)           = X
Number of groups         = 17
Group sample size        = 4
Center =
       X1        X2
61.47059 18.04412
Covariance matrix =
         X1          X2
X1 241.7353 105.38235
X2 105.3824  50.90686
|S| = 1200.545

Control limits:
 LCL       UCL
   0  11.15193
```

Assuming out-of-control points like subgroups 6, 10, and 20 can be avoided in the future, the process should remain in control with the mean vector and covariance matrix approximately equal to the results shown in the summary output above. In this simple analysis of the data from Ryan[83], there were only $m = 20$ subgroups of data. In Chapter 4, Section 4.2, it was recommended that at least 25 subgroups of data be used in a Phase I study in order to get accurate estimates of the in-control process mean and standard deviation when one quality characteristic is charted. When there are p quality characteristics being charted, the number of subgroups required for a Phase I control chart should be even larger.

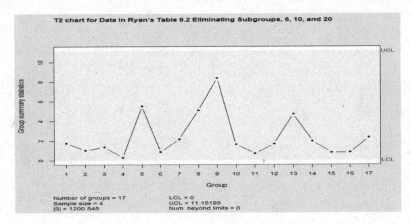

FIGURE 7.6: Phase I T^2 Chart Eliminating Subgroups 6, 10, and 20

7.3.2 Multivariate Control Charts for Monitoring Variability with Sub-grouped Data.

When multivariate data is collected in subgroups, it is possible to monitor the process variability as well as the process mean vector. The covariance matrix **S**, which contains the sample variances for the measured quality characteristics on the diagonals and the sample covariances between measured characteristics on the off-diagonals, is the multivariate analog of the variance s^2 and is the multivariate measure of process variability.

High values for some quality characteristics may be an indicator of a deterioration in quality. However, monitoring the mean vector of quality characteristics is not the only way to detect this possible problem. Increases in the variability correspond to short term spikes that could again indicate sporadic reductions in quality. For that reason, similar to univariate s or R control charts, the process variability indicated by **S** should be monitored in addition to the mean vector $\bar{\mathbf{x}}$.

Alt[2] proposed two different control charts for monitoring the covariance matrix **S**. The simplest of the two is based on the *generalized variance* , $|\mathbf{S}|$ the determinant of the covariance matrix. This is a univariate statistic and Alt constructed a control chart based on the interval $E(|\mathbf{S}|) \pm 3\sqrt{V(|\mathbf{S}|)}$. The expected value and variance of $|\mathbf{S}|$ are:

$$E(|\mathbf{S}|) = b_1|\mathbf{\Sigma}|,$$

and

$$V(|\mathbf{S}|) = b_2|\mathbf{\Sigma}|,$$

where

$$b_1 = \frac{1}{(n-1)^p} \prod_{i=1}^{p} (n-i),$$

and

$$b_2 = \frac{1}{(n-1)^{2p}} \prod_{i=1}^{p} (n-i) \left[\prod_{j=1}^{p} (n-j+2) - \prod_{j=1}^{p} (n-j) \right].$$

The center line and control limits for the control chart are given by

$$UCL = |\mathbf{\Sigma}| \left(b_1 + 3b_2^{1/2} \right)$$
$$CL = b_1 |\mathbf{\Sigma}|$$
$$LCL = |\mathbf{\Sigma}| \left(b_1 - 3b_2^{1/2} \right),$$

and the generalized variances $|\mathbf{S}_i|$ computed within each of the i subgroups are plotted on the chart.

The R package `IAcsSPCR` contains a function `GVcontrol()` for constructing a control chart of the generalized variances. The example code below shows how it can be used to create a control chart of the generalized variances for each subgroup in the data from Ryan's Table 9.2 that was described in the last section. The first argument to the `GVcontrol()` function is the dataframe. In this case the data does not have to be reformatted, although the function expects the data to have consecutively ordered subgroup numbers and an equal number of observations within each subgroup. Figure 7.7 shows the control chart, and the summary statistics the function returns are shown below the code.

```
R>library(IAcsSPCR)
R>data(Ryan92)
R>GVcontrol(Ryan92,20,4,2)
$name
[1] "UCL="

$value
[1] 10771.1

$name
[1] "Covariance matrix="

$value
          x1              x2
```

```
x1 222.0333 103.11667
x2 103.1167  56.57917


$name
[1] "Generalized Variance |S|"


$value
[1] 1929.414


$name
[1] "mean vector="


$value
     x1      x2
60.3750 18.4875


$name
[1] "Subgroup Generalized Variances="


$value
 [1]    45.0555556 2035.6666667 1195.0555556   30.8888889
      9445.5000000   57.0555556
 [7]     4.0000000  452.8333333    1.1111111 3150.1666667
       798.7777778  286.6111111
[13]   453.5000000  101.5000000  120.5555556   47.0555556
         0.3888889   72.5000000
[19]   156.2777778    1.8888889
```

In this control chart, it can be seen that the generalized variances from each subgroup fall within the control limits, and process variability appears to be in control. The data in the dataframe Ryan92 had $p = 2$ quality characteristics and subgroups of size $n = 4$. In order to have accurate estimates of the covariance matrices within each subgroup, the subgroup size n should be substantially larger than the number of quality characteristics p.

In Phase I studies with sub-grouped data, the variance estimates within each subgroup are pooled, or averaged, to get an estimate of the process σ or Σ (for multivariate charts) that becomes the standard for Phase II monitoring. The estimate of the variance is used to judge the magnitude of changes in the mean or mean vector. Recall, that for univariate charts using variables data, it was recommended that the R-chart or s-chart be examined before the $\bar{X}$-chart. If there are assignable causes on the R or s-chart that indicate increases in the variance, those subgroups of data should be removed before constructing the $\bar{X}$-chart. If not, the the control limits on the $\bar{X}$-chart will be too wide and the chart will be less sensitive in detecting changes in the process mean.

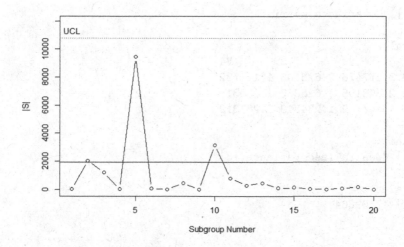

FIGURE 7.7: Control chart of Generalized Variances |S| for Ryan's Table 9.2

An analogous situation occurs with multivariate data. If there are assignable causes in some subgroups that indicate an increase in the covariance matrix, Σ, those subgroups should be removed before finalizing the estimate of Σ and checking for out of control signals on the T^2 chart. The control chart of generalized variances can be used for multivariate data in place of the R or s-chart. As an example of this, consider the data in the dataframe **Frame** that is included in the **IAcsSPCR** package. This dataframe contains $m = 10$ subgroups of data with $p = 3$ quality characteristics. The subgroup size is $n = 10$.

The code below retrieves the dataframe and calls the **GVcontrol()** function to make a control chart of the generalized variances. The control chart is shown in Figure 7.8.

```
R>library(IAcsSPCR)
R>data(Frame)
R>GVcontrol(Frame,10,10,3)
$name
[1] "UCL="

$value
[1] 67.46235

$name
```

```
[1] "Covariance matrix="

$value
          V2         V3         V4
V2 3.477476 2.623105 1.147732
V3 2.623105 4.625853 1.234318
V4 1.147732 1.234318 2.287319

$name
[1] "Generalized Variance |S|"

$value
[1] 17.09664




$name
[1] "mean vector="

$value
     V2       V3       V4
10.4409 17.3352  9.2669

$name
[1] "Subgroup Generalized Variances="

$value
  [1]    6.710274    3.124969    2.327437    6.346057   16.971770
         6.491100    1.070446
  [8]   14.251040    4.186463 112.781321
```

In Figure 7.8, it can be seen that the generalized variance for subgroup 10 falls above the upper control limit. The code below illustrates reformatting the data in Frame so that the mqcc() function can be used to produce a T^2 chart of the data. The T^2 control chart of this data is shown in Figure 7.9.

```
R>library(qcc)
R># reformat as a list of matricies required by mqcc
R>X1<-matrix(Frame$V2,nrow=10,byrow=TRUE)
R>X2<-matrix(Frame$V3,nrow=10,byrow=TRUE)
R>X3<-matrix(Frame$V4,nrow=10,byrow=TRUE)
R>X = list(X1 = X1, X2 = X2, X3=X3) # a list of matrices, one
      for each variable
R>q<-mqcc(X,type="T2",add.stats=TRUE,
    title="T2 Generated data")
```

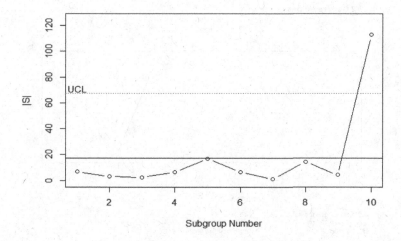

FIGURE 7.8: Control Chart of Generalized Variances |S| for dataframe Frame.

```
R>summary(q)
-- Multivariate Quality Control Chart -------------

Chart type               = T2
Data (phase I)           = X
Number of groups         = 10
Group sample size        = 10
Center =
      X1       X2       X3
10.4409  17.3352   9.2669
Covariance matrix =
          X1        X2         X3
X1 3.477476  2.623105  1.147732
X2 2.623105  4.625853  1.234318
X3 1.147732  1.234318  2.287319
|S| = 17.09664


Control limits:
  LCL       UCL
    0  11.56047
```

The T^2 control chart in Figure 7.9 appears to indicate that there are no assignable causes of differences in the subgroup mean vectors.

However, assuming that whatever caused the increased variances for subgroup 10 can be eliminated in the future, this subgroup should be removed and the two control charts reconstructed.

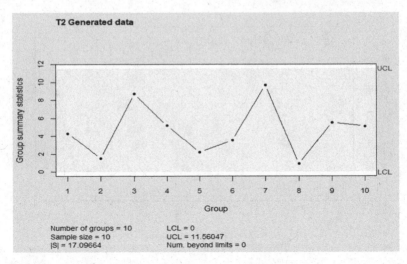

FIGURE 7.9: T^2 Control Chart of Subgroup Mean Vectors in dataframe Frame

The code below removes subgroup 10 and calls the `GVcontrol()` function to make a control chart of the generalized variances for the first 9 subgroups.

```
R>sFrame<-subset(Frame,subgroup!=10)
R>GVcontrol(sFrame,9,10,3)
$name
[1] "UCL="

$value
[1] 35.73629

$name
[1] "Covariance matrix="

$value
          V2        V3        V4
V2 3.003067 2.794676 1.293888
V3 2.794676 4.337586 1.382692
V4 1.293888 1.382692 2.312179

$name
[1] "Generalized Variance |S|"

$value
[1] 9.056467
```

```
$name
[1] "mean vector="

$value
        V2         V3         V4
10.492556  17.338444   9.182111

$name
[1] "Subgroup Generalized Variances="

$value
[1]   6.710274   3.124969   2.327437   6.346057  16.971770
      6.491100   1.070446  14.251040   4.186463
```

The resulting control chart is shown in Figure 7.10. This chart shows the variability now to be in a state of control, and the printed results in the code box above shows that the generalized variance |**S**| has decreased from 17.10 to 9.06.

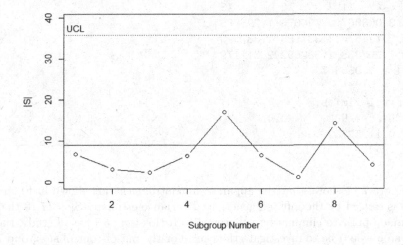

Subgroup Number

FIGURE 7.10: Control chart of Generalized Variances |S| for dataframe Frame Eliminating Subgroup 10.

The code below re-formats the data in the reduced dataframe sFrame and calls the mqcc() function to produce a T^2 control chart. The resulting chart is shown in Figure 7.11.

```
R>library(qcc)
R># reformat as a list of matricies required by mqcc
R>X1<-matrix(sFrame$V2,nrow=9,byrow=TRUE)
R>X2<-matrix(sFrame$V3,nrow=9,byrow=TRUE)
R>X3<-matrix(sFrame$V4,nrow=9,byrow=TRUE)
R>X = list(X1 = X1, X2 = X2, X3=X3) # a list of matrices, one
        for each variable
R>q<-mqcc(X,type="T2",add.stats=TRUE,
   title="T2 Generated data")
R>summary(q)
-- Multivariate Quality Control Chart ------------

Chart type                = T2
Data (phase I)            = X
Number of groups          = 9
Group sample size         = 10
Center =
        X1        X2        X3
10.492556 17.338444  9.182111
Covariance matrix =
         X1        X2        X3
X1 3.003067 2.794676 1.293888
X2 2.794676 4.337586 1.382692
X3 1.293888 1.382692 2.312179
|S| = 9.056467

Control limits:
  LCL      UCL
    0 11.52832
```

Figure 7.11 shows that the apparent in-control situation shown in Figure 7.9 was caused by the inflated generalized variance estimate $|S| = 17.10$ that was made prior to eliminating subgroup 10. If this were a Phase I study, the next step would be to investigate the cause for the out-of-control subgroup 7 in Figure 7.11.

7.3.3   Phase II T^2 Control Chart with Sub-grouped Data

Continuing with the example from Section 7.3.1, suppose new process data was obtained in real-time monitoring. The dataframe Xnew in the R package IAcsSPCR contains 20 new subgroups of 4. The R code below creates a control chart using the Phase II control limits based on the mean and covariance

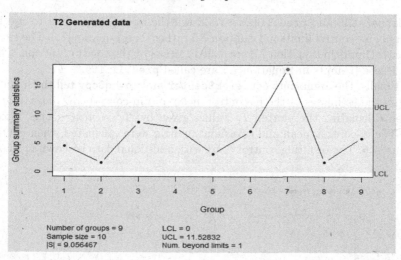

FIGURE 7.11: T^2 Control Chart of Subgroup Mean Vectors in dataframe Frame after Eliminating Subgroup 10.

matrix estimates from the Phase I study after eliminating subgroups 6, 10, and 20, then the mqcc() function adds the new data on the right side of the control chart.

```
R>library(IAcsSPCR)
R>data(Xnew)
R># the next three statements format the new data for the
   mqcc function
R>X1<-matrix(Xnew$x1,nrow=20,byrow=TRUE)
R>X2<-matrix(Xnew$x2,nrow=20,byrow=TRUE)
R>Xn = list(X1 = X1, X2 = X2) # a list of matrices, one for
          each variable
R>library(qcc)
R>qn = mqcc(XR3, type = "T2", newdata=Xn,add.stats=TRUE,
          limits=FALSE,pred.limits=TRUE,center=q3$center,
          cov=q3$cov,title="T2 chart for Phase II data")
```

In the mqcc()} function call, the first argument XR3 tells the function to calculate the control limits based on the data in XR3 that was used in making Figure 7.6. That data contained $m=17$ subgroups. The argument newdata=Xn tells the function to add the reformatted new data on the right. The arguments

limits=FALSE and pred.limits=TRUE tell the mqcc()} function to use the
formula for control limits in Equation 7.5 rather than Equation 7.3. The Phase
I control limits in Equation 7.3 are called limits by the mqcc() function, while
the Phase II limits in Equation 7.5 are called pred.limits.

Finally, the arguments center=q$center and cov=q$cov tell the function
to use the estimates of the in-control mean and in-control covariance matrix
when calculating the plotted T_i^2 values given by: $T_i^2 = n(\bar{\mathbf{x}}_i - \bar{\bar{\mathbf{x}}})'\mathbf{S}^{-1}(\bar{\mathbf{x}}_i - \bar{\bar{\mathbf{x}}})$. The in-control mean and covariance matrix were calculated when making
Figure 7.6. The resulting control chart with additional data is shown in Figure
7.12.

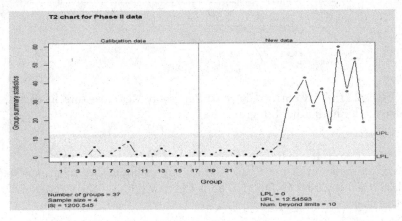

FIGURE 7.12: Phase II T^2 Chart with 20 New Subgroups of Data

Figure 7.12 shows shows that the process mean values for the 11th through
20th new subgroups have T^2 values that fall above the upper control limit.
The UPL=12.54592 can be calculated using Equation 7.5 when $p = 2$, $m = 17$,
and $n = 4$ by modifying the code above Figure 7.2.

There is no need to wait for 20 new subgroups of data to construct the
Phase II control chart. It can be constructed after the first new subgroup by
creating a dataframe Xnew that contained just one subgroup of data, then
reformatting as shown in the code above. The new dataframe Xnew could be
sequentially expanded adding each additional subgroup, then re-creating the
control chart. At each of the sequential steps, any change in the mean vector
would be identified immediately.

7.4 Multivariate Control Charts with Individual Data

7.4.1 Phase I T^2 with Individual Data

Gonzales de la Parra and Rodriguez Loaiza[32] presented data on the impurity profile of a crystalline drug substance determined by high-performance liquid chromatography. Internal reference standards were available for each impurity level reported in the paper. A Phase I study was conducted with single observations on each of 30 lots. The dataframe DrugI in the IAcsSPCR package contains a subset of this data. The code below retrieves the data and uses the mqcc() function to make a T^2 chart for individuals with this data.

```
R>library(IAcsSPCR)
R>qi<-mqcc(DrugI[,-1],type="T2.single",add.stats=TRUE,
        title="Historical data for Drug impurities")
R>summary(qi)
-- Multivariate Quality Control Chart ------------

Chart type              = T2.single
Data (phase I)          = DrugI[, -1]
Number of groups        = 30
Group sample size       = 1
Center =
          A          B          D          E          G
   23.33333   74.00000 625.66667 151.00000 684.00000
Covariance matrix =
            A          B          D          E          G
A   105.74713  -155.1724   294.2529   -51.72414  -551.7241
B  -155.17241  4314.4828 -1399.3103  1813.10345  9424.8276
D   294.25287 -1399.3103 20459.8851  2645.86207 -2857.9310
E   -51.72414  1813.1034  2645.8621  9767.93103 10851.0345
G  -551.72414  9424.8276 -2857.9310 10851.03448 67217.9310
|S|  = 2.893802e+18


Control limits:
  LCL       UCL
    0 12.11333
```

In this code the argument type="T2.single" specifies that this is a T^2 chart for individuals. The output below the code shows the mean vector (over the 30 observations) for impurities A, B, D, E, G, and the covariance matrix calculated with the 30 observations.

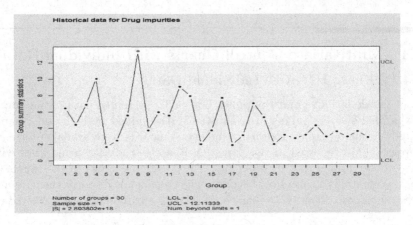

FIGURE 7.13: Phase I T^2 Chart of Drug Impurities

The upper control limit shown in Figure 7.13 and printed below the graph was calculated using Equation 7.7. That calculation can be reproduced with the code below.

```
R># upper control limit calculated with equation 7.7
R>p<-5
R>m<-30
R>n<-1
R>alpha<-1-(1-.0027)^p
R>UCL<-(((m-1)^2)/m)*qbeta(alpha,p/2,(m-p-1)/2,ncp=0,
   lower.tail=FALSE)
R>UCL

[1] 12.11333
```

The control chart in Figure 7.13 shows the eighth individual observation to be out of the control limits. After investigating the cause of the out-of-control observation and updating the OCAP, the control chart should be recreated eliminating observation 8. The code below eliminates the 8th observation and then recreates the control chart as shown in Figure 7.14.

```
R># Redo eliminating observation 8
R>DrugIe<-DrugI[-8,]
R>qi<-mqcc(DrugIe[,-1],type="T2.single",add.stats=TRUE,
        title="Historical data for Drug impurities")
R>summary(qi)
-- Multivariate Quality Control Chart ------------

Chart type               = T2.single
Data (phase I)           = DrugIe[, -1]
Number of groups         = 29
Group sample size        = 1
Center =
        A         B         D         E         G
 23.44828  75.17241 630.00000 155.51724 665.86207
Covariance matrix =
           A          B          D          E          G
A  109.11330  -164.9015   289.2857   -69.70443  -506.6502
B -164.90148  4425.8621 -1607.1429  1713.30049 10422.1675
D  289.28571 -1607.1429 20607.1429  2132.14286  -517.8571
E  -69.70443  1713.3005  2132.1429  9482.75862 13784.3596
G -506.65025 10422.1675  -517.8571 13784.35961 59396.5517
|S| = 1.79408e+18

Control limits:
 LCL      UCL
   0 12.03526
```

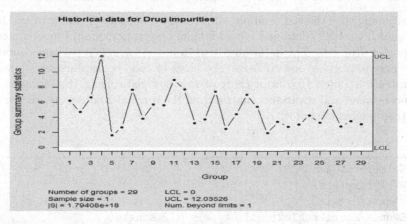

FIGURE 7.14: Phase I T^2 Chart of Drug Impurities after Eliminating Observation 8

In Figure 7.14, it can be seen that the 4th observation falls above the upper control limit. Again, after investigating the cause, this observation should be eliminated and the control chart reconstructed. This reconstructed control chart is shown in Figure 7.15.

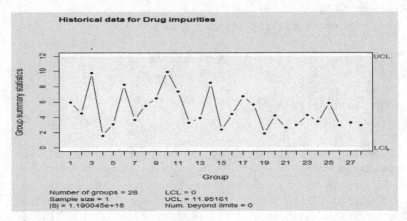

FIGURE 7.15: Phase I T^2 Chart of Drug Impurities after Eliminating Observation 8 and 4

Since there are no points out-of-control in Figure 7.15, it would be appropriate to use the data that was used to construct Figure 7.15 to estimate the process mean vector and covariance matrix.

7.4.2   Phase II T^2 Control Chart with Individual Data

Continuing with the last example, the article by Gonzales de la Parra and Rodriguez Loaiza[32] contained 160 additional observations from Phase II monitoring of drug lots. The dataframe `DrugIn` in the `IAcsSPCR` package contains the first 10 of these observations. The block of code below illustrates how to construct a Phase II control chart of this new data using the estimates of process mean and covariance matrix from the original 20 observations (eliminating numbers 4 and 8).

```
R>library(IAcsSPCR)
R>library(qcc)
R>data(DrugIn)
R>qn = mqcc(DrugIe2[,-1], type = "T2.single",
         newdata=DrugIn[,-1],add.stats=TRUE,
         limits=FALSE,pred.limits=TRUE,center=qi$center,
         cov=qi$cov,
```

```
                title="T2 Individual chart for Phase II data")
-- Multivariate Quality Control Chart ------------

Chart type                    = T2.single
Data (phase I)                = DrugIe2[, -1]
Number of groups              = 28
Group sample size             = 1
Center =
          A          B          D          E          G
   23.92857   71.78571  638.57143  151.78571  668.92857
Covariance matrix =
            A          B          D          E          G
A   106.21693  -122.0899   176.1905  -18.38624   -569.709
B  -122.08995  4244.8413  -793.6508  1396.69312  11120.503
D   176.19048  -793.6508 19160.8466  3173.01587 -1327.513
E   -18.38624  1396.6931  3173.0159  9415.21164  14639.021
G  -569.70899 11120.5026 -1327.5132 14639.02116  61313.624
|S| = 1.190045e+18

New data (phase II)           = DrugIn[, -1]
Number of groups              = 10
Group sample size             = 1

Prediction limits:
 LPL       UPL
   0  22.43832
```

The resulting control chart is shown in Figure 7.16.

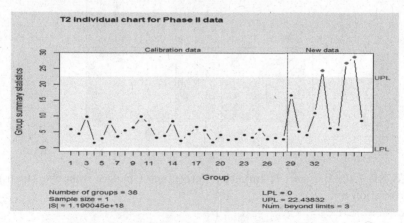

FIGURE 7.16: Phase II T^2 Chart of Drug Impurities

In Figure 7.16, it can be seen that an assignable cause is signaled at the

5th observation in the new data. The upper control limit called (UPL) by the
mqcc() function can be calculated with Equation 7.8 with $m = 28$ and $p = 5$
and again the new data can be added to the chart one observation at a time.

7.4.3   Interpreting Out-of-control Signals

When an out-of-control signal appears on a T^2 control chart, the question may
be asked which of the p quality characteristics (or subset of them) is mainly
responsible for the change. When there are only two quality characteristics like
the example in section 7.2.2.1, the ellipse plot made by the ellipseChart()
function can help. For example, subgroup 10 was above the upper control limit
in Figure 7.2. The T^2 value shown on the chart does not give any information
about which quality characteristic contributes most to the high T^2 value.
However, the ellipse plot, shown in Figure 7.3, indicates that the value of x1
in subgroup 10 is much further from the mean value of x1 than the value of x2
in subgroup 10 is from the mean value of x2. Therefore, the change in quality
characteristic x1 from its mean contributes more to the high T^2 value than
the change in x2.

   When there are more than two quality characteristics, the ellipse plot cannot be made. In that case, one simple way to judge which quality characteristic
(or subset of them) contributes most to the high T^2 value is by making a barchart of the the percentage change of each of the quality control characteristic
from its mean value. For example, Figure 7.17 is a plot of the out-of-control
observation 8 shown in the T^2 chart in Figure 7.13.

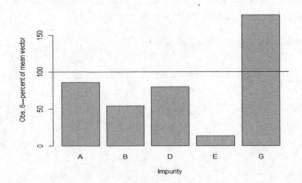

FIGURE 7.17: Barchart of Impurity Percentage Changes from the Mean for
Observation 8

   Here it can be seen that impurities E and G are furthest from their mean
values and therefore changes in these two impurities from their mean probably
contribute most to the high T^2 value for observation 8. However, a simple plot

like this does not take into account the variance and covariances of the five quality characteristics (or impurities).

Runger, Alt and Montgomery[82] discussed the decomposition of the T^2 statistic for an out-of-control point to better determine the contribution of each quality characteristic. This takes into account the variances and covariances of the p quality characteristics. They propose calculating

$$d_i = T^2 - T^2_{(i)} \tag{7.10}$$

where T^2 is the T^2 value of an out-of-control point that is given by Equation 7.2, and $T^2_{(i)}$ is the T^2 for the same observation omitting quality characteristic i and only using the other $p - 1$ quality characteristics in the equation.

Runger, Alt and Montgomery[82] recommend computing the values d_i ($i=$ 1, 2, ..., p) and then focusing on the variables for which the d_i values are relatively large. This can be done using the mqcc() function by running it p times, each time removing one of the p quality characteristics. The T^2 values for each of the observations (or subgroups) can then be retrieved as the statistics component in the object created by mqcc(). For example, when the function call

```
qi<-mqcc(DrugI[,-1],type="T2.single",add.stats=TRUE,
    title="Historical data for Drug impurities")
```
was issued above the T^2 values for the 30 individual observations are stored in the vector qi$statistics.

7.4.4 Multivariate EWMA Charts with Individual Data

When μ_0 and Σ are known, the T^2 chart

$$T_i^2 = n(\mathbf{x_i} - \mu_0)'\Sigma^{-1}(\mathbf{x_i} - \mu_0) \sim \chi_p^2$$

is based on the most recent observation, $\mathbf{x_i}$. Therefore, it is insensitive to detecting small shifts in the mean vector μ_0. Lowry et. al.[65] developed a Multivariate EWMA T^2 control chart (called a MEWMA) that has greater sensitivity for detecting small changes in μ_0. They explained that the natural extension of the univarite EWMA,

$$z_i = \lambda x_i + (1 - \lambda)z_{i-1}$$

is:

$$\mathbf{z_i} = \mathbf{R}(\mathbf{x_i} - \mu_0) + (\mathbf{I} - \mathbf{R})\mathbf{z_{i-1}}. \tag{7.11}$$

where $\mathbf{R} = \text{diag}(r_1, r_2, \ldots, r_p)$ and $0 < r_i \leq 1$, $i = 1, \ldots, p$.

If there is no apriori reason to weigh past observations differently for the p quality characteristics, then $r_1 = r_2 = \ldots = r_p = r$, and $\mathbf{z_i}$ simplifies to:

$$\mathbf{z_i} = r(\mathbf{x_i} - \mu_0) + (1 - r)\mathbf{z_{i-1}}. \tag{7.12}$$

The covariance matrix of z_i is

$$\Sigma_{z_i} = (r/(2-r))\Sigma \tag{7.13}$$

where Σ is the covariance matrix of x_i. An out-of-control signal for the MEWMA occurs when

$$T_i^2 = z_i'\Sigma_{z_i}^{-1}z_i > h_4 \tag{7.14}$$

Lowry et. al.[65] used simulation to generate the upper control limit h_4 that would result in and $ARL_0 \approx 200$. Table 7.1 contains their simulated limits for MEWMA charts with $r = 0.1$.

TABLE 7.1: Upper control limits h_4 for MEWMA Charts with $r = .1$

p	h_4
2	8.66
3	10.79
4	12.73
5	14.56
10	22.67
20	37.01

To illustrate the use of the MEWMA, consider the the following situation. A random sample of 10 observations from a multivariate normal distribution is generated with an in-control mean of $\mu_0 = \begin{pmatrix} 25 \\ 10 \\ 17 \\ 15 \end{pmatrix}$ and a known covariance matrix.

$$\Sigma_0 = \begin{bmatrix} 5.40000000 & 0.09583333 & 2.0583333 & 3.1291667 \\ 0.09583333 & 0.48400000 & 0.2963333 & 0.2686667 \\ 2.05833330 & 0.29633330 & 2.2993333 & 1.0056667 \\ 3.12916667 & 0.26866667 & 1.0056667 & 2.2310000 \end{bmatrix}.$$

Next, another random sample of 15 observations is generated from a multivariate normal distribution with the same covariance matrix, but the mean is shifted to $\mu = \mu_0 + \mu_\Delta$, where $\mu_\Delta = \begin{pmatrix} 1.10 \\ 0.60 \\ 1.00 \\ 0.70 \end{pmatrix}$. The non-centrality factor for the shift in the mean vector is

$$\lambda = (\mu_\Delta)'\Sigma^{-1}(\mu_\Delta) = 1.045907.$$

The code below generates this data and then appends the first sample with the second sample below it to create the matrix D.

```
R># generate random sample of 10  multivariate obs. with p=4
    and mean vector muv
R>muv<-c(25,10,17,15)
R>covm<-matrix(c(5.40000000, .09583333, 2.0583333, 3.12916667,
               .09583333, .48400000,  .2963333, .26866667,
               2.05833333, .29633333, 2.2993333, 1.00566667,
               3.12916667, .26866667, 1.0056667,
               2.23100000),nrow=4,ncol=4 )

R># generate a random sample of 10 obs. with p=4
    and mean vector muv and covariance matrix covm
R>library(MASS)
R>set.seed(100)
R>d1<-mvrnorm(10,mu=muv,Sigma=covm)
R># generate random sample of 15 multivariate obs. with p=4
   and mean vector muv+mud and
R>#  covariance matrix covm
R>mud<-c(1.1,.6,1.0,.70)
R>d2<-mvrnorm(15,mu=muv+mud,Sigma=covm)
R># calculate the noncentrality parameter for detecting the
    mean shift
R>lambda<-mud%*%solve(covm)%*%mud
R>lambda
        [,1]
[1,] 1.045907

R>#combine the data in the combined data the mean shifts after
   the 10th obs.
R>D<-rbind(d1,d2)
R>D
          [,1]       [,2]       [,3]       [,4]
 [1,] 23.75806 10.180360 16.36806 14.52721
 [2,] 25.80020  9.864865 16.93187 14.51697
 [3,] 24.80108  9.839530 17.16286 14.75048
 [4,] 27.39987  9.449689 17.14419 16.24186
 [5,] 24.99657 10.362772 17.00062 15.67339
 [6,] 25.58139 10.273492 17.35958 15.64218
 [7,] 23.42115 10.496906 16.78113 14.27465
 [8,] 26.90254  9.919296 17.15489 15.99281
 [9,] 22.76137 11.021166 17.02295 13.69190
[10,] 25.04470  9.537140 14.01456 15.06273
[11,] 26.22139 10.081466 16.84394 15.80500
```

```
[12,] 28.70585 11.749270 20.93433 17.44007
[13,] 22.40994 10.921784 14.03205 14.18671
[14,] 28.58755  9.707084 17.06339 16.06210
[15,] 24.16904 11.976770 18.30165 15.44996
[16,] 28.96276 10.687066 19.70108 17.54689
[17,] 25.23258 11.534847 17.54417 15.23675
[18,] 28.92701 10.482076 19.86350 17.48245
[19,] 26.96799  9.232787 15.32614 16.64555
[20,] 21.26249 11.550983 15.85880 14.05896
[21,] 25.49466 10.611174 18.18041 13.90000
[22,] 22.14921 10.138899 15.55521 13.76772
[23,] 26.32006 11.504986 17.87188 16.42251
[24,] 29.77446 11.810255 20.04525 19.27867
[25,] 20.84429  9.678784 15.86871 12.55590
```

The R package IAcsSPCR that contains the data sets and functions from this book contains the function MEWMA() that calculates Lowry et. al.'s[65] MEWMA control chart. This function uses a small value of $r = 0.1$ in order to be sensitive to small shifts in the mean vector.

The code below illustrates the use of this function on the matrix of data D created above. The first argument can either be a matrix or dataframe with one row for each observation and one column for each of the p quality characteristics being monitored. This function works for $2 \leq p \leq 10$ quality characteristics. In this example, it is assumed that Σ_0=covm is known and it is supplied to the function as the second argument. The in-control mean vector muv is supplied as the third argument, and the argument Sigma.known=TRUE tells the function that these are the assumed known parameters. When Sigma.known=FALSE then the function calculates mu and covm from the data in D, and they don't need to be supplied to the function.

```
R># MEWMA chart assuming mean and covariance matrix is known
R>library(IAcsSPCR)
R>MEWMA(D,covm,muv,Sigma.known=TRUE)
$name
[1] "UCL="

$value
[1] 12.73

$name
[1] "Covariance matrix="
```

```
$value
              [,1]         [,2]        [,3]        [,4]
[1,]  5.40000000  0.09583333  2.0583333  3.1291667
[2,]  0.09583333  0.48400000  0.2963333  0.2686667
[3,]  2.05833330  0.29633330  2.2993333  1.0056667
[4,]  3.12916667  0.26866667  1.0056667  2.2310000

$name
[1] "mean vector"

$value
[1] 25 10 17 15
```

The results are shown below the commands and Figure 7.18 shows the MEWMA control chart.

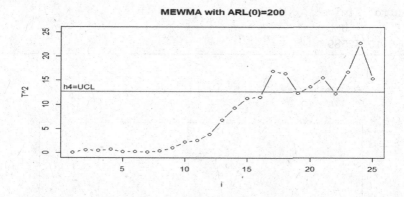

FIGURE 7.18: MEWMA chart of Randomly Generated Data D

The first out of control point occurs at the 17th observation. Since the mean shifted after the 10th observation, it took the MEWMA chart 7 observations to detect this change.

In the code below, the T^2 chart is used on the same data. Again the results are shown below the code and the T^2 control chart is shown in Figure 7.19.

```
R>library(qcc)
R>q <- mqcc(D, type = "T2.single", center=muv,cov=covm,
   confidence.level = (1-.0027)^4)
R>summary(q)
```

```
-- Multivariate Quality Control Chart ------------

Chart type               = T2.single
Data (phase I)           = D
Number of groups         = 25
Group sample size        = 1
Center =
V1 V2 V3 V4
25 10 17 15
Covariance matrix =
              V1           V2           V3           V4
V1 5.40000000 0.09583333 2.0583333 3.1291667
V2 0.09583333 0.48400000 0.2963333 0.2686667
V3 2.05833330 0.29633330 2.2993333 1.0056667
V4 3.12916667 0.26866667 1.0056667 2.2310000
|S| = 0.9210777


Control limits:
 LCL      UCL
   0 10.72589
```

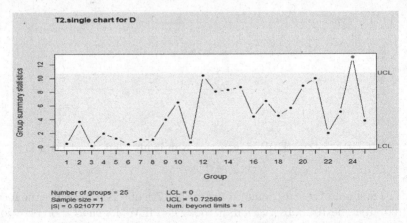

FIGURE 7.19: T^2 Chart of of Randomly Generated Data D

The first out of control signal on the T^2 Chart is at the 24th observation, and it took 7 additional observations to detect the shift in the mean vector.

Lowry et. al.[65] compared the ARL performance of the MEWMA control chart to the T^2 control chart for detecting mean shifts of various sizes. They assumed the mean shift had occurred prior to the application of the chart and made steady state ARL comparisons. Their results are shown in Table 7.2.

TABLE 7.2: ARL Comparisons between T^2 and MEWMA Charts–note: Source Lowry et. al. *Technometrics* 34:46-53, 1992

λ	T^2	MEWMA	T^2	MEWMA	T^2	MEWMA
	$p=2$	$p=2$	$p=3$	$p=3$	$p=4$	$p=4$
0.0	200	200	201	200	200	201
0.5	116	28.1	130	31.8	138	34.7
1.0	42.0	10.2	52.6	11.3	60.9	12.1
1.5	15.8	6.12	20.5	6.69	24.6	7.23
2.0	6.9	4.41	8.8	4.86	10.6	5.18
2.5	3.5	3.51	4.4	3.83	5.19	4.10
3.0	2.2	2.92	2.6	3.2	2.83	3.41
λ	T^2	MEWMA	T^2	MEWMA	T^2	MEWMA
	$p=5$	$p=5$	$p=10$	$p=10$	$p=20$	$p=20$
0.0	200	200	200	200	200	200
0.5	145	31.8	162	48.1	174	64.1
1.0	68.1	11.3	92.8	15.9	117	20.0
1.5	28.5	6.69	44.7	9.16	66.2	11.3
2.0	12.4	4.86	20.6	6.55	34.3	8.00
2.5	5.99	3.83	9.90	5.15	17.4	6.26
3.0	3.31	3.20	5.20	4.28	9.1	5.19

In this table, it can be seen that the MEWMA has shorter ARL's for noncentrality factors $\lambda \leq 2.5$. In the example above the noncentrality factor was $\lambda = 1.0459$, and although the MEWMA chart signaled the out-of-control condition 7 observations sooner than the T^2 chart, on the average it would signal about $60.9-12.1=48.8$ observations sooner, as can be seen by examining the ARLs in the row for $\lambda = 1.0$, and the columns where $p = 4$.

When the in-control process mean vector μ_0 and the covariance matrix Σ are unknown, they must be estimated from a Phase I study. The MEWMA control chart should be constructed using historical data, and out-of-control points should be eliminated before obtaining the estimates of μ_0 and Σ. As in all the previous examples, this is usually an iterative process where the estimates are refined and information is obtained for the OCAP.

When using the MEWMA function to create a Phase II control chart, using the estimated mean vector and covariance matrix from a Phase I control chart, the Sigma.known=TRUE option should be used and the estimates supplied to the function as if they were known values. In other examples in this book where Phase I estimated parameters are used in a Phase II control chart, the control limits are usually widened. For the MEWMA function there is no automatic way of increasing the h_4 upper control limit. This would have to be done manually using trial-and-error experience.

7.5   Summary

This chapter has presented multivariate control charts that should be used when several quality characteristics are being monitored simultaneously. T^2 charts are more sensitive in detecting abnormal or out-of-control conditions than many simultaneous univariate $\bar{X}$-control charts. Likewise, the control chart of the generalized variance will be more sensitive than simultaneous s or R-charts for detecting an increase in variability. The MEWMA (multivariate EWMA chart) will be even more sensitive in detecting small changes in the mean vector than the T^2 charts.

7.6 Exercises

1. Run all the sample code from this chapter.

2. Construct separate univariate control charts for the variables x1 and x2 in the dataframe Ryan92 in the R package IAcsSPCR that contains the data sets and functions from this book.

    (a) Do the results of these two control charts confirm the results the T^2 control charts in Section 7.3.1?

    (b) Explain why or why not.

3. Open the dataframe Sample contained in the R package IAcsSPCR. Determine how many variables (p), how many subgroups (m), and how many observations in each subgroup (n).

    (a) Make a control chart of the generalized variances for each subgroup.

    (b) Is the process variabiliy stable? If not, eliminate any subgroups whose generalized variances are above the UCL.

4. Make a T^2 control chart of the subgrouped data in the dataframe Sample, after having removed any subgroups with generalized variances above the UCL.

    (a) Eliminate any out of control points and reconstruct the control chart.

    (b) Do this iteratively until the process appears to be in control.

5. Make a T^2 chart for individuals using the data in the data frame boiler in the R package qcc.

    (a) Calculate the column means for the in-control points and the column means for the out-of-control points and compare them to the overall column means for the data.

    (b) Using the comparison you made in (a) can you tell what changes in the temperature readings (or columns) are causing the out of control points.

6. The data used to illustrate the MEWMA control chart in Lowry et. al.[65] is available as the dataframe Lowry in the R package IAcsSPCR that contains the data sets and functions from this book. It contains 10 observations with $p=2$ quality characteristics.

(a) Use the MEWMA() function to construct an MEWMA control chart of that data. The data was generated from a bi-variate normal distribution with unit variances and a correlation coefficient of .5. The process mean was on target for at (0,0) for the first 5 observations and shifted to (1,2) for the last five observations. Supply the known covariance matrix and the in-control mean vector to the function.

(b) Does the MEWMA control chart detect the shift in the mean vector? If so how many observations did it take?

(c) Calculate the non-centrality parameter for detecting the shift in the mean vector (Equation 7.9).

(d) Use the mqcc() function in the qcc package to make a Phase I T^2 chart of the same data. Again supply the known covariance matrix and in-control mean to the function as known values.

(e) Does the T^2 Chart detect the shift in the mean vector? If so, how many observations did it take? Based on the non-centrality factor you calculated in (c) is this result surprising?

8

Quality Management Systems

8.1 Introduction

Deming recognized that he was speaking to the wrong audience, when talking to Japanese engineers around 1950. Although the use of statistical methods to improve and control quality had been proven to be effective, Deming felt that their use would wane in Japan, as it had in the US, unless the top executives of companies were convinced of the competitive advantage for using them. In 1950, efforts to improve quality were based on the assumption that defects were costly. However, executives who were used to making decisions on facts had little evidence of that.

Joseph Juran sought to give some evidence in his 1951 *Quality Control Handbook*[51]. The first chapter discussed the economics of quality, stating that there were avoidable and unavoidable costs of achieving a given level of quality. The unavoidable costs were associated with prevention, inspection, sampling, statistical process control, process improvements, and other quality initiatives. The avoidable costs were the costs associated with defects and product failures, such as scrapped materials, labor hours for rework and repair, complaint processing, and financial losses resulting from lost sales.

The last category was hard to quantify, but the other categories could be quantified by the accounting department. Juran could show that increased investment in the unavoidable costs could actually reduce the avoidable costs to more than offset the increase in unavoidable costs. Consequently, the total cost of producing a given quality level could be reduced even without considering the loss of customers and reputation that could result from poor quality. Juran's message gave executives a reason to invest in quality. The financial benefits could be substantial, and Juran referred to improvements in quality as "gold in the mine".

In 1956, Feigenbaum[26] proposed ways that executives could lead the way to "mine the gold". In his 1956 article, and later in a book by the same name, he proposed "Total Quality Control"[27]. The premise was that high quality could not be achieved by leaving responsibility to one group, such as engineering or production. He emphasized that quality had to be everyone's job, led by the top management of the company.

Along with total quality control as a way to "mine the gold" or achieve high quality, it was realized by many that the goal of high quality was a moving target. Competitors were improving quality as well. This became very apparent during the 1980s when many companies lost market share to Japanese competitors. Therefore, the goal changed from simply achieving high quality to continuous improvement in quality. In order to sustain these continuing efforts, top executive leadership became even more necessary.

Feigenbaum noted that all quality control and improvement efforts are involved in the following three activities:

- incoming material control,

- design of the process, and

- process control and improvements.

A useful tool for thinking about these activities is the SIPOC diagram shown in Section 5.1. Specific tasks involved in the three activities are supply chain management, product and process design and development, manufacturing control, and customer relations and customer satisfaction. These tasks involve more than statistical process control and require involvement and cooperation of several functional areas. Therefore, top executives must become involved in coordinating the efforts of different functional areas, setting the standards and defining the important quality measures. In order to coordinate the efforts of different functional areas, a systems approach was needed.

Total Quality Management (TQM) developed out of the philosophies of Deming and Juran in the early 1980s. It was a quality management strategy or system for implementing and guiding quality improvement efforts throughout a company. This strategy sought to incorporate things such as:

- employee involvement and workplace culture

- supplier quality improvement

- increased customer focus,

into the normal business practices in order to facilitate quality improvement efforts. TQM was a popular management practice in the 1980s and had moderate success. The way it was implemented varied from company to company, and some of the reasons for only moderate success were the lack of management focus on reducing variability and use of statistical methods, and the focus on general rather than company-specific objectives[14].

8.2   Quality Systems Standards and Guidelines

A Quality Management System or QMS is based on Juran's Trilogy of Planning, Control and Improvement. The plans for acheiving these objectives is documented in what is called the *quality manual*. Burke and Silvestrini[14] represent a generic quality manual as a four-tiered document known as the document pyramid, as illustrated in Figure 8.1

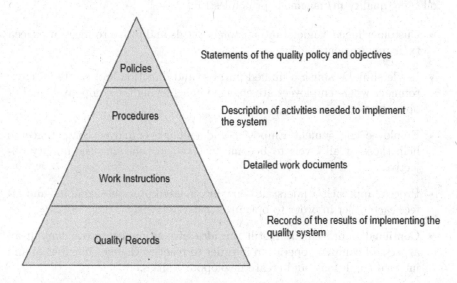

FIGURE 8.1: Quality Manual Documentation Tiers

Quality philosophies and QMSs have revolutionized the business world. To be competitive in markets, adopting a quality management system is no longer an option[12]. The emphasis on continuous improvement in quality and customer satisfaction demands standards and guidelines to assure effective quality management systems.

In previous chapters, ANSI/ASQ/ISO guidelines for sampling inspection, use of control charts, and experimental design methods have been discussed. The next section describes some of the standards that have been developed for entire Quality Management Systems (QMS).

8.2.1   ISO 9000

ISO is an independent, non-governmental network of national standards organizations, and there is only one organization in each country that is a member of ISO. In the US, ANSI (the American National Standards Institute) is the only ISO member. Despite the large number of languages represented among all the international standards organizations, ISO reflects the English language abbreviation of International Standardization Organization and it remains the same in every country. As markets for companies expanded around the world, international standards have become a necessity since there had been differences among national standards. In the 1970s and 80s, a major cooperative effort between countries and international companies resulted in the ISO 9000 family of standards, which is the set of requirements a company must satisfy to become ISO 9000 registered or certified. ISO 9000 is based on the seven following quality management principles[14]:

- Customer focus–Understand customer needs and strive to meet or exceed their requirements.

- Leadership–Establish a unified purpose and direction and create an environment where employees are engaged in achieving the companies' quality objectives.

- Employee engagement–Empower and enhance employee competence to help those at all levels to become involved in achieving the quality objectives.

- Process approach–Understand how interrelated processes combine and affect each other in order to optimize the system.

- Continual improvement–Instill the idea of continual improvement in all aspects of company operation in order to react to changes in external and internal conditions, and create new opportunities.

- Evidence based decision making–Facts, evidence and data analysis provide increased objectivity and confidence in decision making.

- Relationship management–Suppliers, customers, employees, investors and society as a whole influence performance. Therefore, relationships between these groups need to be overseen.

The purpose of the ISO 9000 family is to assist organizations of all types and sizes to develop an effective quality management system[14].

This family of standards consists of:

- ISO 9000: This document provides fundamentals and vocabulary for describing Quality Management Systems. It discusses the component concepts and principles.

- ISO 9001: Information in this document is used in contracts between supplier and customer companies and is often referred to as *external quality assurance*. It provides guidelines that, when followed, will provide confidence to the purchaser that the supplier's quality system is capable of producing the stated quality requirements. It is comprehensive and covers design/development, production, installation, and servicing.

- ISO 9002: This is essentially a subset of ISO 9001, dealing with production and installation. It can be used in contracts with purchasers that do not require standards as stringent as those in ISO 9001.

- ISO 9003: This is the least stringent subset of ISO 9001, dealing only with final inspection and test.

- ISO 9004: This document is a guide to organizations in quality management and is often referred to as *internal quality assurance*. It provides guidelines for all technical administrative and human factors, affecting the quality of products or services at all stages from design to customer satisfaction. It suggests ways to improve organizational performance and customer satisfaction, beyond the ISO 9001 requirements. ISO 9004 are guidelines as opposed to the requirements in ISO 9001–9003. If a company wants to upgrade to a more powerful quality management system, they can begin to selectively apply ISO 9004 guidelines.

ISO first completed the international standards ISO 9000–9004 in 1987. The American National Standards Institute (ANSI) and the American Society for Quality (ASQ) developed ANSI/ASQ Q9000-Q9004. They are technically equivalent to the ISO 9000-9004 with the incorporation of American English usage and spelling to help US companies comply with the requirements. The ISO standards are reviewed and revised, if necessary, every five years. The latest was ISO 9000:2015.

Many organizations become certified as ISO 9001 compliant, and ISO 9001 is the only document in the ISO 9000 family for which an organization can be certified. To become certified, an independent certification organization must audit a company to ensure that its processes are in line with ISO requirements.

Figure 8.2 illustrates the major steps in the ISO 9001:2015 certification-registration process. In the initial step, the company or organization seeking to be registered interviews several potential registrars and selects one. The second step involves filling out the questionnaire and associated paperwork provided by the registrar. The next step is to sign a contract with the registrar and confirm the schedule. Next, a first-stage assessment provides the opportunity to identify areas of non-compliance early on. In the document review, the registrar compares the companies' quality manual to the ISO 9001 stated procedures and requirements. In the certification assessment, the registrar observes the companies processes to make sure the policies and procedures in the quality manual are being followed. If there are major non-compliances, in the reviews, the company has time to make corrective action before the registration process is complete.

How long it takes a company to become certified depends upon how much the company already practices ISO 9000 principles. It could take anywhere from 6 months to 18 months to become certified.

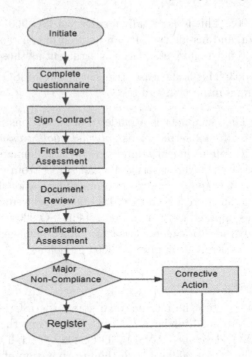

FIGURE 8.2: Flowchart of ISO 9000 Registration Process

Some of the potential benefits of ISO 9000 certification are:

1. Increased business efficiency

2. Higher customer satisfaction

3. Lower unnecessary costs

4. More reliable quality of processes and products

5. New customer opportunities.

New customer opportunities arise in various ways. The EC/EFTA (European Commission and Europe Free Trade Association) either requires or strongly encourages ISO 9000 certification. Therefore, ISO 9000 certification opens this marketplace to registered companies. In the US, the Food and Drug Administration Center for Medical Devices and Radiological Health replaced their Good Manufacturing Practice Guidelines by ISO 9000. The US Department of Defense has also adopted the equivalent ANSI/ASQ Q9000. However, most companies are driven to seek ISO 9000 certification more by contract negotiations than by requirements. For example, a company involed in EFTA trade may be required to seek ISO 9000 certification. They, in turn may require all their suppliers (through contracts) to also seek ISO 9000 certification as well. Thus, supply chain management practices also limit potential suppliers to be ISO registered companies.

8.2.2 Industry Specific Standards

Some industries have developed their own industry-specific standards that emulate ISO 9000. For the Aerospace industry there is AS 9100, QS9000 for the automotive industry, and TL 9000 for the telecommunications industry. QS9000 was later revised by the International Automotive Task Force as IATF 1649:2016, and is intended to be used in conjuction with ISO 9001:2015 to define requirements for automotive products[14].

There have been some criticisms of ISO 9000 and the industry-specific standards in the past. Montgomery[72] stated that certification audits focused too much attention on checking policy documentation and not enough attention on verifying the perceived benefits 1 to 4 in the list shown in the last section. He cites the nearly 300 deaths caused by Brigestone/Firestone tire failures resulting in rollover accidents of Ford Explorer vehicles. The Brigestone/Firestone plant was QS9000 certified, yet the accidents led to a recall of over 6.5 million tires. However, with the 5 year revision cycle for the ISO standards, changes are made in response to valid critique and problems like this are less likely to occur in the future.

8.2.3 Malcolm Baldridge National Quality Award Criteria

In 1987, by an act of the US congress, the Malcolm Baldrdge National Quality Award (MBNQA) was established. This award was patterned after the Deming Prize in Japan. The purpose of the award was to recognize quality achievement in individual companies, share successful quality strategies, and promote awareness among US businesses. The award was named after the late Secretary of Commerce Malcolm 'Mac' Baldrige who had supported the creation of such an award as a way to improve the United States competitive position. His service as Secretary of Commerce was one of the longest in history, and he was one of the most beloved and well-respected secretaries who held that position[90].

The Baldrige Award was more than a set of guidelines for a QMS, and its broader purpose was to help the entire United States improve quality and productivity by the following actions[90]:

- Stimulating companies to attain excellence for the pride of achievement.

- Recognizing outstanding companies to provide examples to others

- Establishing guidelines that business, governmental, and other organizations can use to evaluate and improve their own quality efforts.

- Providing information from winning companies on how to manage for superior quality.

The legislation that created the Baldridge award gave the US Department of Commerce responsibility for it, and the Commerce Department gave the role of administering the award to the National Institute of Standards and Technology (NIST), an agency within the Department of Commerce. Figure 8.3 shows the award program organization.

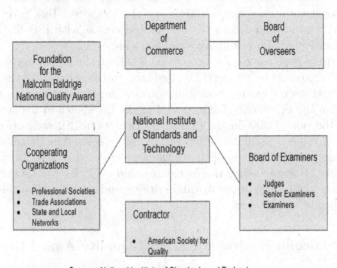

Source: National Institute of Standards and Technology

FIGURE 8.3: Malcolm Baldrige Quality Award Organization

The Board of Overseers, consisting of at least five people who are renowned in the quality management field, monitors the effectiveness of the Award. Members are appointed to the Board by the Secretary of Commerce, and they also review the work of the contractor, the American Society for Quality.

The board of examiners consists of Judges, Senior Examiners, and Examiners. To become a board member, an application must be made, and selections are based on qualifications, experience, and recognition of expertise. Selections are made annually. Examiners do not represent any one professional group or company, but they are selected from various fields such as manufacturing, service, academia, and trade associations. NIST appoints senior examiners, and the judges from among the examiners. Each year the examiners are required to participate an a three-day training program during which the award criteria and scoring system are reviewed with case studies.

The contractor, the American Society for Quality participates in criteria development, publicity, administering the award ceremony, and organizes an annual Quest for Excellence Conference featuring Award winners.

The legislation that created the Award did not appropriate government funds for its operation. However, in addition to the support from the Department of Commerce and NIST, the private sector shows its support by raising money, sharing quality information and volunteering. The Foundation for the Malcolm Baldrige National Quality Award was set up to create an endowment, which collects the application fees and supports the award program.

The Baldrige award[75] was originally given in three categories, but over the past 31 years, it has expanded to the six categories shown below:

- manufacturing

- service

- small business

- education

- health care

- non-profit

A total of 18 awards can be given in any given year, with no limit on the individual categories. However, in some past years no award has been given in some categories.

Steeples[90] explained that the Baldrige Award has become the standard for excellence in quality management. Award winners can be assured that they can deliver goods and services as well as the best organizations in the world. The QMS essentials are embodied in the Baldrige criteria:

- Customers define quality

- Senior company or organization leadership must create a clear quality values and build them in to normal operating procedures

- Excellent quality evolves from well-designed and executed systems and processes

- Continuous improvement must be integrated into all processes.

- Organizations must develop goals and strategic operational plans to achieve quality leadership

- Shortened response time for all operations and processes must be part of quality improvement efforts.

- Operations and decisions of the organization must be based on facts

- All employees must be appropriately trained, developed, and involved in quality improvement efforts

- Design quality and error prevention must be key elements

- Organizations must communicate quality requirements to suppliers and work with them to elevate their efforts.

The Baldrige criteria for performance excellence is a structure that any organization can use to improve its overall performance[75]. The assessment criteria consist of seven categories:

1. **Leadership** – examines how senior executives lead the organization and how the organization handles its responsibilities to the public and to the environment in which it is inserted.

2. **Strategic planning** – examines how an organization sets strategic directions and how it determines key plans of action.

3. **Customers** – examines how an organization determines requirements and expectations of customers and markets, how it builds relations with customers, and how it acquires, satisfies, and retains customers.

4. **Measurement, analysis, and knowledge management** – examines management, effective use, analysis and enhancement of data and information to provide support for key processes at the organization and for the organization's performance management system.

5. **Workforce** – examines how an organization allows its workforce to develop their full potential and how the workforce is aligned with the organization's objectives.

6. **Operations** – examines aspects of how key production/delivery and support processes are designed, managed, and enhanced.

7. **Results** – examines an organization's performance and improvement in its main business areas: customer satisfaction, financial and market performance, human resources, performance of suppliers and partners, operational performance, governance, and social responsibility. This category also looks at an organization's performance in relation to its competitors.

There are at least two items considered by the examiners on each assessment criteria (more details can be found at[74]). Points are awarded for each of the categories, but unlike early criticism of ISO 9001 requirements, a large portion of the points are awarded in the **Results** category. Paraphrasing Steeples[90]: Leadership is the driver of the quality bus, strategy is the road map, measurement and knowledge management is the fuel, workforce is the engine, operations are the landmarks and results are the destination.

The MBNQA criteria are dynamic and just like the continuous improvement model that award applicants are encouraged to follow, the MBNQA Board gleans information from the Award program each year and incorporates the best ideas about assessment criteria, review processes, examiner training, and applicant eligibility requirements into the process to be used in the following year.

Although one of the intents of the Baldrige Award was to recognize companies for outstanding achievements in quality, many companies and organizations have sought to utilize the freely available Baldrige assessment criteria to improve without actually competing for an award. Companies doing this have found that they discover their own areas of competence that they can exploit in the marketplace. They also find areas for improvement. Teamwork and better communication is fostered by self assessment and planing improvements. Finally, senior management involvement and commitment is increased by the self assessment. This is because the stringent Baldrige criteria requires alignment of all quality management practices, whereas traditional management approaches have usually assigned responsibility for quality to company departments whose efforts were often hidden from top management[90].

For companies and organizations seeking examples of how to make improvements with regard to the Baldrige assessment criteria, the annual Quest for Excellence® conference is the definitive showcase of best practices from Baldrige Award winners. At this conference presentations are given on best practices and insights on how excellent performance is achieved, and a pre-conference workshop is held for beginner and intermediate users of the Baldrige Criteria for Performance Excellence.

Some may ask if using the Baldrige assesment citeria is really worth the the effort. In an article in 2011 Bailey[5] examined results from the five businesses that were two-time Baldrige Award winners. She found that the median growth in sites was 67% over 6 years, the median growth in revenues was 93%, and the median growth in jobs was 63%. Impressive results!

8.3 Six-Sigma Initiatives

8.3.1 Brief Introduction to Six Sigma

Many new QMSs have been introduced in recent decades. Of those new strategies, the Six Sigma initiative has been a major breakthrough. Using Six Sigma allows a company to consistently meet customer expectations or requirements. Therefore, according to ISO, it speaks the language of business[17].

Madrigal[69] lists the Six-Sigma Building Block Definitions as:

- **Unit:** The item produced, purchased, or experienced.

- **Defect:** An even that does not meet a customer specification

- **Defective:** A unit with one or more defects

- **Defect Opportunity:** A measurable chance for a defect to occur

- **DPMO:** Defects per million opportunities

- **Process Yield:** Proportion of non-defective units produced by the process

- **Process Sigma:** An expression of process yield based on the number of defects per million opportunities

One minus the Process Yield, or proportion outside the specification limits, is illustrated in Figure 8.4[25] and Table 8.1.

FIGURE 8.4: Motorola Six Sigma Concept

TABLE 8.1: Motorola Six Sigma % and ppm out of Spec

Spec Limit	% outside specs	ppm Defective
$\pm 1\sigma$	69.77	697,700.0
$\pm 2\sigma$	30.87	308,700.0
$\pm 3\sigma$	6.68	66,810.0
$\pm 4\sigma$	0.621	6,210.0
$\pm 5\sigma$	0.0233	233.0
$\pm 6\sigma$	0.00034	3.4

If the process is centered within the specification limits with a $C_p = 2.0$, then the closest a specification limit will be to the process mean is 6σ or six process standard deviations. With no shift in the mean, the proportion above the upper specification limit or below the lower specification limit will be `pnorm(-6)=9.865876e-10`. If the mean shifts 1.5 standard deviations to the left, the proportion below the lower specification limit will be `pnorm(-4.5)=3.397673e-06`, and the proportion above the upper specification limit will be `1-pnorm(7.5)=3.18634e-14`, thus the total proportion out of specification limits would be `3.397673e-6+3.18634e-14=3.397673e-06`.

By symmetry, if the process mean shifts to the right 1.5 standard deviations, the proportion outside the specification limits would be the same, i.e., 3.397673e-06=.00034 or 3.4 ppm.

If a company can reduce the process standard deviation to be 1/12 of the difference between the customer desired upper and lower specification limits and prevent the process mean from changing by more that 1.5 standard deviations, they will be able to meet the customer expectation consistently. Based on this fact a description of Six Sigma consistent with its developers is: "An effective application of statistical techniques, delivered in an innovative manner that has achieved acceptance, use, and results by the management of many organizations"[53].

In order to improve quality by reducing variability, the Six Sigma method utilizes what is called the DMAIC process (Define, Measure, Analyze, Improve, and Control), which is a way of looking at the scientific method that is clear to managers and employees. To facilitate the use of the DMAIC process, the roles of all employees and managers within the organization are defined using a comparison to the roles of students and instructors within a martial arts school.

8.3.2 Organizational Structure of a Six Sigma Organization

Continuous improvement in a Six Sigma organization is accomplished on a project-by-project basis. The organization of a Six Sigma organization expedites the selection and completion of the projects that will lead to the greatest improvements. Figure 8.5 shows the structure of a Six Sigma organization[89].

In this figure, the Leadership Executive Team is the executive and his or her direct staff that is responsible for the business unit. This team has the overall responsibility for approving and providing resources to the improvement projects undertaken. The Champion or Project Sponsor is a business leader who identifies areas for possible improvement and brings them to the attention of the Leadership Executive Team. A Black Belt (BB) and team members will undertake work on an approved project. Black Belts are experienced in the Six Sigma tools and train and direct their team members to complete project work. Black Belts typically work with more than one team and one project at any time. The team members are selected from among the Green Belts (GBs), who have less training and expertise in the use of Six Sigma tools but come from cross-functional areas and together have knowledge in all aspects of the process under study. Green belts normally work full time in the process and are selected to be part of a temporary team to solve particular problems. The Master Black Belts (MBBs) are experts in the use of the Six Sigma tools and problem solving. Their job is to help Project Sponsors identify potential projects and train Black Belts and Green Belts in the use of Six Sigma methodology. This is a full time job.

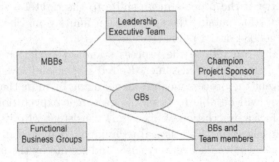

FIGURE 8.5: Six Sigma Organization

8.3.3   The DMAIC Process and Six Sigma Tools

Improvement projects are completed in a Six Sigma organization using the
DMAIC process. The steps in the DMAIC process, as steps in the PDCA
cycle, are similar to the steps in the scientific method. Figure 8.6 represents
the steps in the DMAIC process.

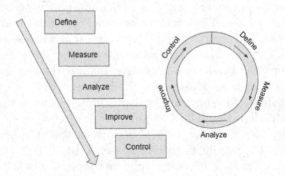

FIGURE 8.6: DMAIC Process

Sometimes an improvement project may be completed in one cycle of the
DMAIC process, as depicted on the left side of Figure 8.6. In other cases it
may take several cycles to complete a project, as represented on the right side
of Figure 8.6. Occasionally, the completion of one project naturally leads to
the beginning of another project. In that situation, the DMAIC process is
again used repeatedly to complete more than one project.

8.3.4 Tools Used in the DMAIC Process

The tools used at each step use of the DMAIC process on improvement projects clarify the relevance of this process to business managers.

8.3.5 DMAIC Process Steps

Define: The purpose of the Define step is to identify a problem or opportunity for improvement. The goal of this step is to provide the evidence needed to reach a consensus decision on whether this project should go forward. In this step, a Project Charter is developed. This charter is like a contract between the project team and the project sponsor or champion. It should specify the goal of the project, the budget and timeline for completing the project, and expected deliverables in terms of:

- Customer satisfaction

- Profits

- Employee involvement etc.

Some additional tools described in this book that are useful at this step are:

- SIPOC Diagram–Chapter 4

- Flowchart–Chapter 4, etc.

Measure: The purpose of this step is to quantify what is currently happening in the process and identify how it has changed or remained constant over time. It necessarily involves a data collection plan. The goal is to develop baseline information that will be later compared to the improved results that will occur. Benchmarking is a good way to quantify current process results and identify areas for improvement at the same time. Benchmarking consists of comparing process results to the results of other processes in other areas of the company or within trusted customer or supplier companies. This practice may reveal gaps between current process results to what may be possible. Other tools described in this book that are useful at this step are:

- Phase I control charts–Chapter 4 and 7

- Histograms and Process Capability Studies–Chapter 4

- Line graphs–Chapter 4

- Acceptance Sampling Plans–Chapters 2 and 3

Analyze: The purpose of this step is to discover root causes and their effect, then to prioritize opportunities for improvement. Tools described in this book that are useful at this step are:

- Flowcharts–Chapter 4

- Brainstorming, Cause and Effect Diagram and Ask Why?–Chapter 4

- Pareto charts–Chapter 4

- PDCA–Chapter 4

- Design and Analysis of Experiments and Hypothesis Testing–Chapter 5

Improve: The purpose of this step is to review the results of the Analyze step and identify the best ideas to improve the process. Next these ideas should be implemented in the process and their effectiveness should be evaluated. Some tools described in this book that are useful at this step are:

- Analysis of data from designed experiments, confirmation experiments, and ranking effects–Chapter 5

- PDCA–Chapter 4

- Phase II Control Chart appended to Phase I Chart with Phase I limits–Chapter 4

Control: The purpose of this step is to sustain the potential gains identified in the Analyze step and ensure that the potential solution identified in the previous step is embedded in the process. Mistake proofing or Poka-yoke should be used to prevent previous errors. Some additional tools described in this book that are useful at this step are:

- Robust Design Experiments–Chapter 5

- Phase II monitoring with control charts–Chapters 6 and 7

- Use of OCAP–Chapter 4

8.3.6 History and Results Achieved by the Six Sigma Initiative

Mikel Harry, who is considered the "Godfather" of the Six Sigma initiative, introduced the idea in the mid 1980s at Motorola, and Motorola formally launched a Six Sigma Program in 1987[16]. Motorola set an ambitious goal of improving all products (goods as well as services) tenfold within 5 years. The Six Sigma program allowed Motorola to focus all resources on reducing variation in both manufacturing processes and administrative processes, and the results of this effort came quickly. From 1987 to 1997 Motorola achieved a fivefold growth in sales, and profits grew nearly 20%. Cumulative savings were $US14 billion and stock prices grew at an annual rate of 21.3 percent[53].

In 1988 Motorola corporation won the Baldrige Award in the manufacturing category. As part of the Baldrige tradition of providing information about how they managed for quality, Motorola prepared an informational package describing their Six Sigma program[70]. Motorola joined forces with companies like IBM, ABB, Texas Instruments, Allied Signal, and Kodak to found the "Six Sigma Research Institute".

Six Sigma achieved high success in the General Electric company in the 1990s, and using it in nearly 30 service businesses (non-manufacturing), General Electric saved $1billion over a two-year period. Much of the current popularity of Six Sigma is due to promotion by the former GE CEO Jack Welch[69]. Other companies soon began to use the Six Sigma approach. For example, AlliedSignal obtained savings of $2billion over a five-year span, and other well known examples of Six Sigma companies include ABB, Lockeed Martin, Poloroid, Sony, Honda, American Express Ford, Lear Corporation and Solectron[53].

8.3.7 Six Sigma Black Belt Certification

Just as companies benefit from using Six Sigma methodology for managing their businesses and seeking continuous improvement, individuals also benefit when they gain knowledge and experience in the use of the Six Sigma method. Requirements for some entry-level positions in companies applying Six Sigma methodology include certification as a Six Sigma Green Belt.

Many organizations now issue certificates or certifications for Six Sigma Green Belts and Six Sigma Black Belts. These organizations fall into the categories of professional societies, universities, companies, and online education for profit organizations. Companies may provide training to their own employees and issue a certificate of completion. However, the two most respected certifications are from The American Society for Quality (ASQ) and the International Association for Six Sigma Certification (IASSC). Of these, the ASQ certifications are considered the Gold Standard and are most respected by employers.

Applicants for ASQ certification are required to take a comprehensive exam over the published "Body of Knowledge". In addition, applicants for Green Belt certification must have 3 years of job experience working on teams using methods described in the ASQ Six Sigma Green Belt Body of Knowledge. Applicants for the Six Sigma Black Belt certification must have led two successful improvement projects teams using the DMAIC process, in addition to passing a written exam. The proof of having led successful projects must be given in the form of signed affidavits from the project sponsor or champion.

Other online certification bodies do not require work experience but require completion of a course and or an exam. ASQ also offers a Quality Process Analyst certification that requires at least two years of college and passing an exam. The exam for this certification is as rigorous as the Black Belt exams given by some online certification bodies, and with successful completion students obtain a certification before graduation that greatly increases their employment opportunities after graduation[42].

The benefits to individuals who become certified in areas of Six Sigma methodology include:

- Improved capabilities as a leader

- Knowledge applicable across many industries

- Increased earnings.

The increased earnings of certified individuals is well documented, and ASQ publishes an annual salary survey in the *Quality Progress* magazine. Six Sigma Black Belts may earn 160% of their uncertified counterparts doing comparable jobs.

8.4 Additional Reading

A step beyond Six-Sigma is the idea of Lean Six Sigma. This is a strategy that incorporates the collaborative teamwork and the DMAIC process from the Six-Sigma initiative with the ideas of lean manufacturing. The purpose is to reduce waste at the same time as reducing variability and improving customer satisfaction. Waste, as described by Fujio Cho of Toyota, is anything other than the minimum amount of equipment, materials, parts, space, and workers time, which are absolutely essential to add value to the product[93].

The eight common types of waste are: Rework, Over Processing, Over Production, Transportation, Waiting, Excess Inventory, and Wasted Motion[68]. Lean Six Sigma is an organized approach to eliminate these wastes. Although Six Sigma and Lean are different, they have many similarities and work well together. Over 350 Fortune 500 Companies use Lean Six Sigma and typical annual savings is on the order of $500 million[68].

Books by George, Rowlands and Kastle[31], Summers[93], and Morgan and Brenig-Jones[73] provide more in-depth information about Lean Six Sigma.

8.5 Summary

During World War II, the use of statistical methods, such as acceptance sampling and statistical process control in US defense contractor plants, were very effective in improving quality, increasing productivity, and reducing costs. The use of these techniques was mandated by the US Department of Defense just prior to and during WW II. When the war ended and companies returned to producing goods for civilian use, there was no mandate to use these statistical methods and they fell into disuse.

Quality gurus such as Deming, Juran, and Feigenbaum argued that use of these statistical methods made good business sense, and that top management needed to be involved to bolster and support their use. When many US companies lost markets and market share during the 1980s, top business executives became interested in quality control and the idea of Quality Management Systems (QMS) emerged. The idea was that proper leadership, customer focus, continual improvement using a process approach, employee involvement, and fact-based decision making would lead to higher customer satisfaction and business success.

By 1980s, national and international guidelines for effective QMSs appeared and ISO began certifying organizations that were deemed compliant to the ISO 9001 requirements for a QMS. In the US, the Baldrige Award criteria went a step further in creating a competition among organizations in following the Baldrige requirements. The purpose was to stimulate improvements in quality and productivity among US organizations. Award winners shared their strategies and very impressive results have been documented.

One Baldrige Award winner, Motorola Corporation, introduced the Six Sigma approach. This approach included an innovative way of including statistical methods in a project-by-project improvement strategy that has achieved world wide acceptance by top company executives. The use of this approach has yielded enormous results in customer satisfaction and business success by the organizations who use it.

The Six Sigma approach is compatible with the ISO 9000 requirements and the Baldrige Award criteria. Due to the organizational structure of a Six Sigma company and the techniques used in the DMAIC process, the use of the statistical tools advocated by Deming and described in this book are guaranteed.

Many institutions certify individuals as Six Sigma Black Belts and Six Sigma Green Belts, and this opens job opportunities and higher salaries to those certified. However, most of the Six Sigma certification exams for individuals focus on multiple choice concept questions and the ability to demonstrate the use of hand calculations and table look ups for applying statistical techniques. Despite the advances in the use of computers and results from recent research, students who want to take certification exams will have to be familiar with these hand calculations and table look ups. The American Society for Quality publishes certification handbooks and sample exam questions to help in preparation for the exams.

The use of computers, and results of recent research, can enable dramatic increases in the efficiency of Statistical Quality Control methods over what was possible in 1950. In Chapters 2 and 3, it has been shown that the use of the computer makes it quick and easy to develop OC curves and average sample size curves that are useful in comparing different acceptance sampling plans. Understanding how to develop and interpret these curves can greatly clarify the differences in various proposed sampling plans. In addition, programs have been written that allow quick access to published sampling schemes that are normally used in domestic and international trade.

The use of the computer greatly reduces the time involved to iteratively develop Shewhart control charts with Phase I data. Other simple computer displays of data, such as Pareto Diagrams and Scatter plots etc., greatly help in determining the assignable causes for out-of-control conditions that may appear on Phase I control charts. More sophisticated statistical tools, such as PDCA and designed experiments, can detect assignable causes that may not be obvious with observational data.

More recently developed time-weighted control charts such as the Cusum and EWMA charts and multivariate control charts, can greatly reduce the expected time to detect out-of-control signals in Phase II monitoring of process data. In addition, they can reduce the chance of falsely concluding an assignable cause is present when there is none. This can eliminate wasted time looking for non-existent problems.

This book has emphasized the use of the public domain program R for SPC and Acceptance Sampling calculations. This open-source software is quickly gaining international acceptance for this purpose. Students who learn to use this tool will be able to make immediate contributions in the job market where there is a strong demand for Quality Analysts, Quality Engineers, Six Sigma Green Belts, and Six Sigma Black Belts.

Bibliography

[1] *NIST/SEMATECH e-Handbook of Statistical Methods.* 2012. https://www.itl.nist.gov/div898/handbook/pmc/section3/pmc321.htm.

[2] F. B. Alt. Multivariate Quality Control. In N. L. Johnson and S. Kotz, editors, *Encylclopedia of Statistical Sciences*, volume 6. John Wiley and Sons, New York, 1985.

[3] J. Anhoej. *qicharts: Quality Improvement Charts*, 2017. R package version 0.5.5, https://CRAN.R-project.org/package=qicharts.

[4] Division ASQ-Statistics. *Glossary and Tables for Statistical Quality Control.* ASQ Quality Press, Milwaukee, Wisconsin, 3rd edition, 1996. ISBN 0-87389-354-9.

[5] D. Bailey. The Verdict is In–Baldrige Is About Revenue and Jobs, 2011. https://www.nist.gov/blogs/blogrige/verdict-baldrige-about- revenue-and-jobs, last accessed 6/18/2020.

[6] S. Bersimis, S. Psarakis, and J. Panaretos. Multivariate Statistical Process Control Charts: An Overview. *Qual. Reliab. Engng. Int.*, 23:517–543., 2007.

[7] C. M. Borror, C. W. Champ, and S. E. Rigdon. Poisson EWMA Control Charts. *Journal of Quality Technology*, 30(4):352–361, 1998.

[8] C.M. Borror and C.W. Champ. Phase I Control Charts for Independent Bernoulli Data. *Quality and Reliability Engineering International*, 17:391–396, 2001.

[9] G. E. P. Box. Use and Abuse of Regression. *Technometrics*, 8(4):625–629, 1966.

[10] G.E. P. Box, S. Bisgaard, and C. Fung. An Explanation and Critique of Taguchi's Contributions to Quality Engineering. *Quality and Reliability Engineering International*, 4(2):121–131, 1988.

[11] G. C. Britz, D. W. Emerling, L. B. Hare, R. W. Hoerl, S. J. Janis, and J. E. Shade. *Improving Performance Through Statistical Thinking.* ASQ Quality Press, Milwaukee, Wisconsin, 2000. ISBN 0-87389-467-7.

[12] B. Brocka and M. S. Brocka. *Quality Management- Implementing The Best Ideas of the Masters.* Business One Irwin, address = Homewood, Ill., year = 1992, note = ISBN 1-55623-540-2,.

[13] R.K. Burdick, C. M. Borror, and D. C. Montgomery. *Design and Analysis of Gauge R&R Studies.* SIAM Society for Industrial and Applied Matematics, Philadelphia, PA, 2005. ISBN 0-89871-588-1.

[14] S. E. Burke and R. T. Silvestrini. *The Certified Quality Engineer Handbook 4th Ed.* ASQ Quality Press, Milwaukee, Wisconsin, 2017. ISBN 978-0-87389-944-4.

[15] E. Cano, J. Moguerza, M. Prieto, and A. Redchuk. *SixSigma:Six Sigma Tools for Quality Control and Improvement*, 2012. R package version 0.9-52, https://CRAN.R-project.org/package=SixSigma.

[16] E. L. Cano, J. M. Mogguerza, and M. P. Corcoba. *Quality Control with R - An ISO Standards Approach.* Springer, New York, N.Y, 2015. ISBN 978-3-319-24044-2.

[17] E. L. Cano, J. M. Mogguerza, and A. Redchuck. *Six sigma with R - Statistical Engineering for Process Improvement.* Springer, New York, N.Y, 2012. ISBN 978-1-4614-3651-1.

[18] S. Chakraborti, S. W. Human, and M. A. Graham. Phase I Statistical Process Control Charts: An Overview and Some Results. *Quality Engineering*, 21:52–62, 2009.

[19] C. Christensen, K.M. Betz, and M.S. Stein. *The Certified Quality Process Analyst Handbook.* ASQ Quality Press, Milwaukee, Wisconsin, 2nd edition, 2013. ISBN 978-0873898652.

[20] S.P. Cunningham and J.G. Shanthikumar. Empirical Results on the Relationship Between Die Yield and Cycle Time in Semiconductor Wafer Fabrication. *IEEE Transactions on Semiconductor Manufacturing*, 9(2):273–277, 1996.

[21] W.E. Deming. *Out of the Crisis.* MIT Center for Advance Engineering Study, Cambridge, Mass., 1986. ISBN 0-911379-01-0.

[22] H.F. Dodge and H.G. Romig. Single and Double Sampling Inspection Tables. *Bell System Technical Journal*, 20(1):1–61, 1941.

[23] H.F. Dodge and H.G. Romig. *Sampling Inspection Tables, Single and Double Sampling.* John Wiley and Sons, New York, 2nd edition, 1959.

[24] B. Durakovic. Design of Experiments Application, Concepts, Examples: State of the Art. *Periodicals of Engineering and Natural Sciences*, 5(3):421–439, 2017.

[25] T. English. Six sigma handbook for the modern engineer, 2020. https://interestingengineering.com/the-six-sigma-handbook-for-the- modern-engineer.

[26] A. V. Feigenbaum. Total quality control. *Harvard Business Review*, Nov.-Dec.:94–98., 1956.

[27] A. V. Feigenbaum. *Total Quality Control*. McGraw-Hill, New York, N.Y, 1961.

[28] H. F. Freeman. Statistical Methods for Quality Control. *Mechanical Engineering*, page 261, 1937.

[29] A. Gandy and J. Kvaloy. *spcadjust:Functions for Calibrating Control Charts*, 2015. R package version 1.1, https://CRAN.R-project.org/package=spcadjust.

[30] A. Gandy and J. T. Kvaloy. Guarranteed Conditional Performance of Control Charts via Bootstrap Methods. *Scandinavian Journal of Statistics*, 40:647–668, 2013.

[31] M. George, D. Rowlands, and B. Kastle. *What is Lean Six Sigma?* McGraw-Hill Education, New York, N.Y., 2003. ISBN 978-0071426688.

[32] M. Gonzales de la Parra and P. Rodriguez-Loaiza. Application of the Multivariate T^2 Control Chart and the Mason-Tracy-Young Decomposition Procedure to the Study of the Consistency of Inpurity Profiles of Drug Substances. *Quality Engineering*, 16:127–142, 2003-04.

[33] U. Groemping. *DoE.base: Full Factorials, Orthogonal Arrays and Base Utilities for DoE*, 2019. R package version 1.1-3, https://CRAN.R-project.org/package=DoE.base.

[34] U. Groemping. *FrF2: Fractional Factorial Designs with 2-Level Factors*, 2019. R package version 2.1, https://CRAN.R-project.org/package=FrF2.

[35] M. Hamada and C. F. J Wu. Analysis of Experiments with Complex Aliasing. *Journal of Quality Technology*, 24:130–137, 1992.

[36] D. M. Hawkins. A Cusum for a Scale Parameter. *Journal of Quality Technology*, 13(4):228–231, 1981.

[37] D. M. Hawkins. Cusum control charting: An underutilized spc tool. *Quality Engineering*, 5(3):463–477, 1993.

[38] D. M. Hawkins and D. H. Olwell. *Cumulative Sum Charts and Charting for Quality Improvement*. Springer, New York, N.Y., 1997. ISBN 0-387-98365-1.

[39] W. J. Hill and W. R. Demier. More on Planning Experiments to Increase Research Efficiency. *Industrial and Engineering Chemistry*, 62(10), 1970.

[40] K. Holtzblatt, J. B. Wendell, and S. Wood. *Rapid Contextual Design*. Elsevier Inc., San Francisco, CA, 2005. 978-0-12-354051-5.

[41] Youden W. J. and E. H. Steiner. *Statistical Manual of the Association of Official Analytical Chemists*. Association of Official Analytical Chemists, 1975.

[42] J. Jacobsen. Competitive Edge for New Grads, 2009. https:// lawson-jsl7.netlify.app/CQPAedgeforGrads.pdf, last accessed 6/24/2020.

[43] L. Johnson and K. McNeilly. Results May Not Vary. *Quality Progress*, 44(5):61–65, 2011.

[44] M. E. Johnson and M. Boulanger. Statistical Standards and ISO, Part 4 Applications Related to the Implementation of Six Sigma. *Quality Engineering*, 24:552–557, 2012.

[45] Mark Johnson and Bradley Jones. Classical Design Structure of Orthogonal Designs with Six to Eight Factors and Sixteen Runs. *Quality and Reliability Engineering International*, 27:61–70, 2010.

[46] B. L. Joiner. *Fourth Generation Management - The New Business Consciousness*. McGraw-Hill, Inc., New York, NY, 1994. ISBN 0-07-032715-7.

[47] B. Jones and D. C. Montgomery. Alternatives to Resolution IV Screening Designs in 16 Runs. *Int. J. Experimental Design and Process Optimisation*, 1:285–295, 2010.

[48] B. Jones and C. J. Nachtsheim. A Class of Three-Level Designs for Definitive Screening in the Presence of Second Order Effects. *Journal of Quality Technology*, 43(1):1–15, 2011.

[49] B. Jones and C. J. Nachtsheim. Definitive Screening with Added Two-Level Categorical Factors. *Journal of Quality Technology*, 45(2):121–129, 2013.

[50] B. Jones and C. J. Nachtsheim. Effective Design-Based Model Selection for Definitive Screening Designs. *Technometrics*, 59(3):319–329, 2017.

[51] J Juran. *Quality Control Handbook*. McGraw-Hill, New York, N.Y, 1951.

[52] G. J. Kerns. *Introduction to Probability and Statistics Using R*. G. J. Kerns, 2011. https://archive.org/details/IPSUR.

[53] B. Kiefsjo, H. Wiklund, and R. L. Edgeman. Six Sigma Seen as a Methodology for Total Quality Management. *Measuring Business Excellence*, 5(1):31–35, 2001.

[54] A. Kiermeier. *AcceptanceSampling: Creation and Evaluation of Acceptance Sampling Plans*, 2019. R package version 1.0-6, https://CRAN.R-project.org/package=AcceptanceSampling.

[55] S. Knoth. *spc: Statistical Process Control – Calculation of ARL and Other Control Chart Performance Measures*, 2019. R package version 0.6.3, https://CRAN.R-project.org/package=spc.

[56] T. Kourti and J. F. MacGregor. Multivariate SPC Methods for Process and Product Monitoring. *Journal of Quality Technology*, 28:409–428., 1996.

[57] J. Lawson. *Design and Analysis of Experiments with R*. Chapman and Hall/CRC, Boca Raton, Florida, 2015. ISBN 978–1-4398-6813-3.

[58] J. Lawson. *daewr: Design and Analysis of Experiments with R*, 2016. R package version 1.1-7, https://CRAN.R-project.org/package=daewr.

[59] J. Lawson. *AQLSchemes: AQL based Acceptance Sampling Schemes*, 2020. R package version 1.6-9, https://CRAN.R-project.org/package=AQLSchemes.

[60] J. Lawson. *IAcsSPCR:An Intro. to Accept. Samp. and SPC/R*, 2020. R package version 1.2.1, https://r-forge.r-project.org/.

[61] J. S. Lawson. Phase II Monitoring of Variability using Individual Observations. *Quality Engineering*, 31(3):417–429, 2019.

[62] W. Li and C. Nachtsheim. Model Robust Factorial Designs. *Technometrics*, 42:345–352, 2000.

[63] W. Libbrecht, F. Deruyck, H. Poelman, A. Verberckmoes, J. Thybaut, J. DeClercq, and P. VanDerVoort. Optimiaztion of Soft Templated Mesoporus Carbon Cynthesis using Definitive Screening Design. *Chemical Engineering Journal*, 259:126–134, 2015.

[64] G.J. Lieberman and G.J. Resnikoff. Sampling Plans for Inspection by Variables. *Journal of the American Statistical Association*, 50:467–516, 1955.

[65] C. A. Lowry, W. H. Woodall, C. W. Champ, and S.E. Rigdon. A Multivariate Exponentially Weighted Moving Average Control Chart. *Technometrics*, 34:46–53, 1992.

[66] J.M. Lucas. Counted Data CUSUM's. *Technometrics*, 27(2):129–143, 1985.

[67] J.M. Lucas and R.B. Crosier. Fast Initial Response for CUSUM Quality Control Schemes: Give your CUSUM a Head Start. *Technometrics*, 24(3):199–205, 1982.

[68] J. L. Madrigal. The power of lean/six sigma: How to leverage it? Lecture Notes, 2010. Oxford Worldwide Group.

[69] J. L. Madrigal. Six Sigma Overview. Lecture Notes, 2020. Oxford Worldwide Group.

[70] A. Mitra. *Fundamentals of Quality Control and Improvement*. Prentice Hall, Upper Saddle River, New Jersey, 1998. ISBN 0-13-645086-5.

[71] D.C. Montgomery. *Design and Analysis of Experiments*. John Wiley & Sons, Hoboken, New Jersey, 6th edition, 2005. ISBN 0-471-48735-X.

[72] D.C. Montgomery. *Introduction to Statistical Quality Control*. John Wiley & Sons, Hoboken, New Jersey, 7th edition, 2013. ISBN 978-1-118-14681-1.

[73] J. Morgan and M. Brenig-Jones. *What is Lean Six Sigma?* John Wiley & Sons, New York, N.Y., 2015. ISBN 978-1119067351.

[74] NIST. Baldrige Excellence Builder, 2017-2018. https://www.nist.gov/system/ files/documents/2017/02/09/2017-2018-baldrige-excellence-builder.pdf, last accessed 2020-6-18.

[75] NIST. Baldrige Performance Excellence Program, 2019-2020. https://www. nist.gov/baldrige/publications/baldrige-excellence-framework, last accessed 2020-6-18.

[76] P. S. Olmstead. How to Detect the Type of an Assignable Cause. *Industrial Quality Control*, 9(4):22, 1952.

[77] J. O'Neill, G. Atkins, D. Curbison, B. Flak, J. M. Lucas, D. Metzger, L. Morse, T. Shah, T. Steinmentz, K. VanCitters, M. C. Wiener, B. Wingerd, and A. Yourey. Statistical Engineering to Stabilize Vaccine Supply. *Quality Engineering*, 24(2):227–240, 2012.

[78] E. R. Ott. *Process Quality Control-Troubleshooting and Interpretation of Data*. McGraw-Hill, New York, N. Y., 1975. ISBN 0-07-047923-2.

[79] R. L. Plackett and J. P. Burman. The Design of Optimum Multifactor Experiments. *Biometrika*, 33:305–325, 1946.

[80] L.D. Romboski. *An Investigation of Quick Switching Acceptance Sampling Systems*. PhD thesis, Rutgers-The State University, New Brunswick, N.J., 1969.

[81] T. Roth. *qualityTools: Statistical Methods for Quality Science*, 2016. R package version 1.55, https://CRAN.R-project.org/package=qualityTools.

[82] G. C. Runger, F. B. Alt, and D. C. Montgomery. Contributors to a Multivariate Statistical Process Control Signal. *Communications in Statistics - Theory and Methods*, 25:2203–2213, 1996.

[83] T. P. Ryan. *Statistical Methods for Quality Improvement*. John Wiley and Sons Inc, New York, N.Y., 2011. ISBN 978-0-470-59074-4.

[84] E.G. Schilling and D. V. Neubauer. *Acceptance Sampling in Quality Control*. Chapman and Hall/CRC, Boca Raton, Florida, 3rd edition, 2017. ISBN 978-1498733571.

[85] L. Scrucca. *qcc: Quality Control Charts*, 2017. R package version 2.7, https://CRAN.R-project.org/package=qcc.

[86] W.A. Shewhart. *Economic Control of Quality of Manufactured Product*. D. Van Nostrand, New York, 1931. republished in 1980 by the American Society for Quality, Milwaukee, WI.

[87] Shmueli. *Practical Acceptance Sampling - A Hands-on Guide*. Axelrod Schnall Piblishers, Green Cove Springs, FL, 2rd edition, 2016. ISBN 978-0-991-5766-7-8.

[88] R. D. Snee. Experimenting with a Large Number of Variables. In Ronald Snee, Lynne B. Hare, and J. Richard Trout, editors, *Experiments in Industry Design Analysis and Interpretation of Results*. American Society for Quality Control, Milwaukee, Wisconsin, 1985. ISBN 0-087389-001-9.

[89] R. D. Snee and R. W. Hoerl. *Six Sigma Beyond the Factory Floor*. Pearson Prentice Hall, Upper Saddle River N. J., 2005.

[90] M. M. Steeples. *The Corporate Guide to The Malcolm Baldridge National Quality Award-Proven strategies for building quality into your organization*. ASQ Quality Press and Business One Irwin, Milwaukee WI, and Homewood IL, 2nd edition, 1993. ISBN 1-55623-957-2.

[91] S. H. Steiner. EWMA Control Chart with Time Varying Control Limits and Fast Initial Response. *Journal of Quality Technology*, 31(1):75–86, 1999.

[92] K.S. Stephens and K.E. Larson. An Evaluation of the MIL-STD-105D System of Sampling Plans. *Industrial Quality Control*, 23(7), 1967.

[93] D. Summers. *Lean Six Sigma: Process Improvement Tools and Techniques*. Prentice Hall, Upper Saddle River, N. J., 2011. ISBN 978-0-13-512510-6.

[94] D. C. S. Summers. *Quality*. Prentice Hall, Upper Saddle River, N.J., 2nd edition, 2000. ISBN 0-13-099924-5.

[95] G. Taguchi and Y. Wu. *Introduction to Off-Line Quality Control.* Central Japan Quality Control Association, 1979.

[96] A.I.A.G. TaskSubcommittee. *Statistical Process Control (SPC) Reference Manual.* Automotive Industry Action Group, Troy Michigan, 1992.

[97] L.C. Vance. Average Run Lengths of Cumulative Sum Control Charts for Controlling Normal Means. *Journal of Quality Technology,* 18:189–193, 1986.

[98] M. Walton. *The Deming Management Method.* Dodd Mead and and Company, Inc., New York, N.Y., 1986. ISBN 0-396-08683-7.

[99] G. Wernimount. Ruggedness Evaluation of Test Procedures. *ASTM Standardization News,* 5:61–64, 1977.

[100] C.H. White and J. B. Keats. ARLs and Higher-Order Run-Length Moments for the Poisson CUSUM. *Journal of Quality Technology,* 28(3):363–369, 1996.

[101] C.H. White, J.B. Keats, and J. Stanley. Poisson Cusum versus c-Chart for Defect Data. *Quality Engineering,* 9(4):673–679, 1997.

[102] H. Wickham and G. Grolemund. *R for Data Science-Import, Tidy, Transform, Visualize and Model Data.* O'Reilly Media Inc., Sebastopol, CA, 2017. 978-1-491-91039-9.

Index

Printed in the United States
by Baker & Taylor Publisher Services